T0321014

Plant Nematodes of Agricultural Importance

A Color Handbook

John Bridge

Tropical Plant Nematology Advisor, Emeritus Fellow
CAB International UK Centre, Egham, Surrey, UK

James L. Starr

Professor, Department of Plant Pathology and Microbiology
Texas A&M University, College Station, Texas, USA

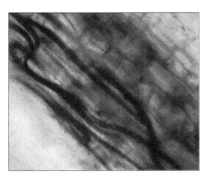

 CRC Press
Taylor & Francis Group
Boca Raton London New York

CRC Press is an imprint of the
Taylor & Francis Group, an **informa** business

CRC Press
Taylor & Francis Group
6000 Broken Sound Parkway NW, Suite 300
Boca Raton, FL 33487-2742

© 2007 by Taylor & Francis Group, LLC
CRC Press is an imprint of Taylor & Francis Group, an Informa business

No claim to original U.S. Government works

Printed on acid-free paper
Version Date: 20161109

International Standard Book Number-13: 978-1-84076-063-7 (Hardback)

Visit the Taylor & Francis Web site at
http://www.taylorandfrancis.com

and the CRC Press Web site at
http://www.crcpress.com

Contents

Preface

Soil and plant nematodes are one of the most numerous groups of organisms occurring in the soil. They are microscopic animals and, with a few exceptions, are not visible to the naked eye. The majority of the soil-borne nematodes are not pests of crops and feed on other organisms, particularly bacteria and fungi. Those that are parasitic on crop plants can be very damaging and, because of their microscopic size, associating them with crop damage is therefore mainly dependent on determining the symptoms of their effects on plants or plant growth. Some of the parasitic nematodes do produce characteristic and recognizable symptoms of damage but many of them only produce nonspecific symptoms. The damage and the symptoms caused can be visible above-ground; however, apart from poor growth and yield of the plants, the specific nematode-induced damage symptoms can often only be seen in the below-ground plant organs, mainly the roots, rhizomes, bulbs, corms, and tubers.

When the above-ground symptoms are the result of nematodes feeding on roots below ground, these symptoms are often similar to those seen when there are water or nutrient deficiencies in the soil. These so-called poor soils or 'tired soils' are often a result of a build-up of large populations of parasitic nematodes in the soil. The main above-ground symptoms of nematode root damage are poor or stunted growth, reduced foliage, twig/branch dieback, chlorosis or yellowing of leaves, poor fruit or seed production and, in extreme cases, wilting and early senescence or death of plants. The below-ground symptoms are related to the health of the root systems and other structures which can be reduced in number, suffer from necrosis and rot, or have abnormal growths such as swellings or galling. Generally, it is necessary to examine roots and other plant tissues to establish a connection between damage symptoms and nematodes. To be certain of the association between particular nematodes, the organisms have to be extracted from the soil, roots, or other plant material and identified microscopically.

This book is written to help people working with plants to have an improved understanding of plant nematode pests and enable them to give better informed diagnoses, where possible, of the damage caused by the very wide range of nematodes that are known to parasitize and injure plants. The colour photographs provide an extensive range both of symptoms of nematode injury and of nematodes themselves, observed in the field or microscopically in plant tissues, throughout the world. There is an introductory chapter on biology and parasitism. Parasitic nematodes are discussed under the crops that they are known to attack. Crop chapters are divided into grain legumes, vegetables, flowers, cereals, root and tuber crops, and tree, plantation, and cash crops. Concise information is provided for the nematodes in these crop sections on their distribution, symptoms and diagnosis, management, and identification. A final chapter outlines the common methods used in nematology.

A glossary of nematological terms and a bibliography of the most important nematology texts and papers have also been produced to help the reader.

In the compilation of this book, we are grateful to our nematological colleagues for their advice, support, and encouragement. We also wish to thank those who have generously allowed us to use their photographs; these persons are acknowledged specifically in the legends of the photographs. Lastly we thank Monica and Marylou, our long-suffering wives, for their patience and support of our efforts.

John Bridge
James L. Starr

CHAPTER 1

Plant Nematode Biology and Parasitism

- INTRODUCTION
- MIGRATORY ECTOPARASITES
  Belonolaimus spp.; *Xiphinema, Longidorus,* and *Paralongidorus* spp.; *Anguina, Aphelenchoides,* and *Ditylenchus* spp.
- MIGRATORY ENDOPARASITES
  Pratylenchus spp.; *Radopholus* spp.; *Hoplolaimus* spp.; *Ditylenchus* spp.

- SEDENTARY ENDOPARASITES
  Meloidogyne spp.; *Nacobbus* spp.; *Globodera* and *Heterodera* spp.; *Rotylenchulus* spp.; *Tylenchulus* sp.

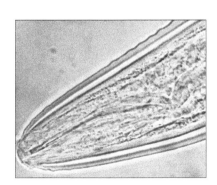
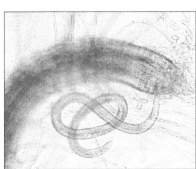

Introduction

Plant parasitic nematodes are principally aquatic animals requiring free moisture for activity; they inhabit the moisture films surrounding soil particles and the moist environment of plant tissues. Nearly all plant parasites spend a portion of their life cycle in the soil. Most nematodes are adapted to the subtropical to tropical climates, but some are adapted to the cooler climates of the more northerly and southerly latitudes or to higher elevations. Any climate that supports higher plants will also support a population of plant parasitic nematodes that are similarly adapted to that climate. Like most invertebrate organisms, their level of activity is closely linked to the environmental conditions, especially temperature.

Most nematodes and the problems they cause are typically associated with coarsely textured soil with relative large pore spaces. Soils with sand contents of greater than 60% fulfill these conditions, as do soils with high concentrations of organic matter or with low sand concentrations but with a high degree of aggregation of silt and clay such that pore spaces are increased. However, substantial nematode damage has been observed in nearly all soils types.

The nematode life cycle is relatively simple, consisting of the egg, four juvenile stages, and the adults. The length of the life cycle varies greatly among the different genera, ranging from a few days to nearly 1 year under optimal environmental conditions and a favourable host. Nematodes reproduce both sexually and by various asexual, parthenogenetic mechanisms. Males are common in species reproducing sexually, but are generally rare or unknown in species reproducing parthenogenetically. Reproductive potential also varies greatly with genera, with some producing less than 50 eggs/female to those that may produce more than 1000 eggs/female.

All crop plants are susceptible to at least one nematode species. Thus the potential exists for nematode parasitism in all climates on any crop. The degree of damage caused by nematodes to the crop in any given field is closely related to the nematode population density, especially for annual crops. Precise data are lacking on the distribution of most agriculturally important nematodes, but the distribution of the cyst nematodes on potato in Europe and on soybean in the central USA, and of root knot and reniform nematodes on cotton in the southern USA is well documented. In each of these cases, there are regions in which more than 50% of the fields are infested with the problem nematode species. In many other fields, even though potentially damaging species are present, their numbers and reproductive potential in that environment is insufficient to cause measurable yield losses. With perennial crops, even low initial numbers of nematodes under conducive conditions can increase sufficiently to cause substantial yield suppression.

Although most nematodes are root parasites, there are nematodes adapted to parasitism in nearly all plant tissues and organs. Plant parasites can be conveniently classified based on their mode of parasitism (**Table 1**). The symptoms of nematode damage vary greatly and may be quite indistinct. It has been documented that with some crops a suppression of yield occurs prior to the expression of diagnostic symptoms.

This text provides general information on the biology and parasitic habits of the most agriculturally important genera of plant parasitic nematodes and specific information on the most important species.

TABLE 1 Parasitic habits and examples of nematode genera

1. Ectoparasites: generally the nematodes remain on the surface of the plant tissues, feeding by inserting the stylet into cells that are within reach
   - Foliar ectoparasites: ectoparasites feeding generally on epidermal plant cells of young leaves, stems, and flower primordia often enclosed by other foliage (*Anguina, Aphelenchoides, Ditylenchus* spp.)
   - Root ectoparasites:
     - • Ectoparasites with short stylets feeding mainly on outer root cells and root hairs (*Tylenchorhynchus, Trichodorus, Paratrichodorus,* some *Helicotylenchus* spp.)
     - • Ectoparasites with long stylets that can be inserted deep into root tissues normally at the growing tip (some can become relatively immobile) (*Belonolaimus, Cacopaurus, Criconemoides, Dolichodorus, Hemicriconemoides, Hemicycliophora, Longidorus, Paralongidorus, Paratylenchus, Xiphinema* spp.)

2. Migratory endoparasites: all stages of the nematodes can completely penetrate the plant tissues, remaining mobile and vermiform and feeding as they move through tissues; they often migrate between the soil and roots
   - Foliar endoparasites: endoparasites in stems, leaves, flower primordia, or seeds (*Aphelenchoides, Bursaphelenchus [Rhadinaphelenchus] cocophilus, Bursaphelenchus xylophilus, Ditylenchus angustus, Ditylenchus dipsaci*)
   - Below-ground endoparasites: all stages of endoparasites are found throughout different tissues in roots, corms, bulbs, tubers, and seeds (peanuts) (*Aphasmatylenchus, Ditylenchus* [some], *Helicotylenchus* [some], *Hirschmanniella, Hoplolaimus, Pratylenchoides, Pratylenchus, Radopholus, Rotylenchus* [some], *Scutellonema* spp.)

3. Sedentary endoparasites: immature female or juvenile nematodes completely enter the plant tissues, develop a permanent feeding site, become immobile, and swell into obese bodies. Expansion of plant tissues (galling) can occur around nematodes (*Achlysiella, Globodera, Heterodera, Meloidogyne, Nacobbus, Punctodera* spp.)

4. Semi-endoparasites: immature female or juvenile nematodes only partially penetrate the roots, leaving the posterior half to two-thirds of the body projecting into the soil. Nematodes become immobile in a fixed feeding site and the projecting posterior of the body becomes enlarged. (Some migratory nematodes can also be found in a semi-endoparasitic position on the roots.) (*Rotylenchulus, Sphaeronema, Trophotylenchulus, Tylenchulus* spp.)

Migratory ectoparasites

This group includes a large number of genera and species. Many ectoparasitic species are only rarely associated with crop damage. Among the migratory ectoparasites most noted as aggressive pathogens (those that typically cause extensive damage to the host) of several crop species are the sting nematode (*Belonolaimus longicaudatus*), the sheath nematode (*Hemicycliophora arenaria*), the dagger nematodes (*Xiphinema* spp.), the needle nematodes (*Longidorus* spp., *Paralongidorus* spp.), and the stubby root nematodes (*Paratrichodorus* and *Trichodorus* spp., 7). Various stunt nematodes (*Tylenchorhynchus* (1), *Quinisulcius*, and *Merlinius* spp.) and ring nematode (*Criconemoides* spp.) (2) are less commonly associated with substantial yield losses but can be economically important on selected crops. Some *Helicotylenchus* (3) and *Rotylenchus* species feed as ectoparasites. All life stages of the migratory ectoparasites, except the egg, are parasitic. Root feeding nematodes with short stylets feed on the epidermal and outer cortical cells often at the root tips (4), those with longer stylets feed deeper in the cortex. Eggs are laid singly in the soil and the life cycle is usually straightforward, with each stage having to feed on a suitable host in order to develop to the next stage (5). Reproduction is by both sexual and asexual systems.

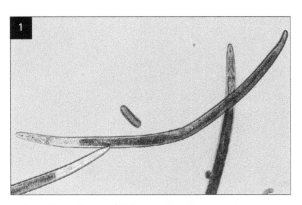

1 Females and egg of *Tylenchorhynchus annulatus*.

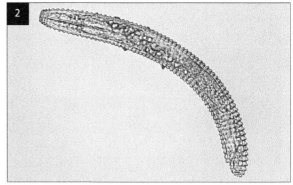

2 *Criconemoides* female.

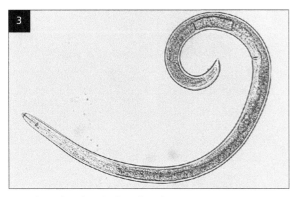

3 *Helicotylenchus indicus* female.

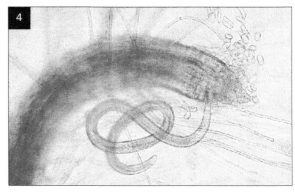

4 *Tylenchorhynchus maximus* feeding as an ectoparasite on a root tip.

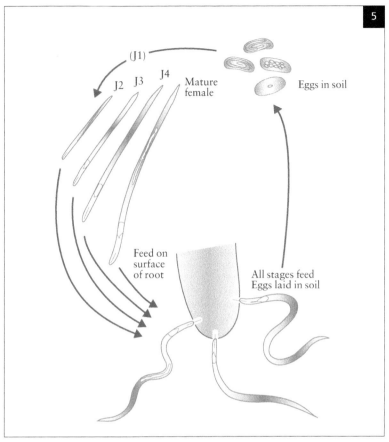

5 Life cycle of the ectoparasite *Longidorus* spp.

Belonolaimus spp.

The sting nematode, *Belonolaimus longicaudatus*, is a migratory ectoparasite with a long feeding stylet (6, 25) that allows it to pierce roots and feed on the cortical tissues several cell layers deep in the roots. The sting nematode does not induce any specialized feeding sites, but causes extensive cellular damage and necrosis of cortical tissues. Distribution of this species is strongly influenced by soil texture and temperatures. The nematode is found only in the soil in warm climates and requires very sandy soils (typically with greater than 85% sand content). During the warmest parts of the year, population densities may decline in the upper soil profile, with most nematodes being found 15–30 cm deep. The reproductive potential of the sting nematode is moderate and rarely do population densities exceed 100 nematodes/100 cm³ soil.

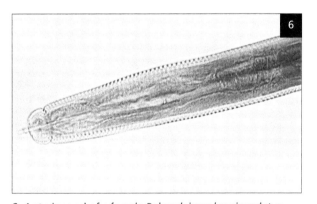

6 Anterior end of a female *Belonolaimus longicaudatus*.

Xiphinema, Longidorus, and Paralongidorus spp.

The dagger (*Xiphinema*) and especially the needle nematodes (*Longidorus* and *Paralongidorus* spp.) (137) are among the largest plant parasitic nematodes. They typically feed at the root tip, with several nematodes often feeding on a single root tip (5). These feeding activities often result in a swollen root tip, such that the symptoms may be confused with root galls caused by the root knot nematodes. Similar to the root knot and other sedentary nematodes, host cells fed upon by *Xiphinema* and *Longidorus* often become enlarged, multinucleate, and with a more dense cytoplasm than adjacent cells. The long stylets of these nematodes permit them to feed on cells in the interior of the root apex while the nematode body is exterior to the root. These nematodes have relatively long life cycles, depending on environmental conditions and susceptibility of the host plant. Females of *Xiphinema* spp. range from 1 to over 10 mm in length. They have life cycles typically requiring 1 year from egg to egg for *X. americanum* and up to 3 years for *X. diversicaudatum*. Several species of these genera are also important because they vector plant viruses, such as the grapevine fanleaf virus by *X. index* and raspberry ringspot virus by *L. elongatus*. This group of nematodes encompasses a large number of different species, the taxonomy of which is dynamic, and accurate identification of species is difficult. They parasitize a large number of vegetable crops, field crops, and perennial fruit crops. Many species are important in the cooler climates of northern Europe and North America.

Anguina, Aphelenchoides, and Ditylenchus spp.

Some species of the genera *Anguina* (90), *Aphelenchoides* (72), and *Ditylenchus* are migratory ectoparasites of the foliage, feeding on stems, leaves, and inflorescences; they can also be found endoparasitically in some plant structures. *Anguina tritici* is the infamous ear cockle nematode of wheat; the two most destructive and widely distributed species of *Aphelenchoides* are *A. besseyi*, the rice white tip nematode, and *A. ritzemabosi*, the chrysanthemum leaf nematode. *A. ritzemabosi* invades leaves through the stomata to feed endoparasitically on mesophyll cells; otherwise they are ectoparasites on the surface of leaves and buds. Eggs are laid in the leaf tissues and the life cycle is very short taking only 10 days. All stages are vermiform and migratory. *Ditylenchus* species often have both ectoparasitic and endoparasitic feeding phases on their hosts (page 13).

Aphelenchoides are relatively long thin nematodes similar in many respects to *Ditylenchus*. The *A. ritzemabosi* female is around 1 mm in length with a posterior vulva and long tapering tail. Stylets are small, 12 μm long, with tiny basal knobs. Males are present and morphologically similar to the female apart from sexual organs.

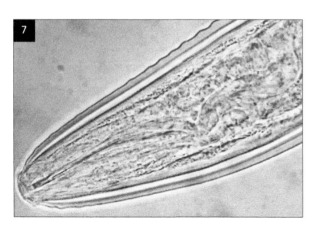

7 Head of *Trichodorus* sp. showing curved stylet.

Migratory endoparasites

Many of the important parasites of plants are migratory endoparasites that remain vermiform throughout their lives. The species well documented as causing crop damage mainly belong to the genera *Pratylenchus*, *Radopholus*, *Hirschmanniella* (**124, 125**), *Hoplolaimus*, *Scutellonema* (**113**), *Aphasmatylenchus*, and *Ditylenchus*, and also some *Helicotylenchus* (**3**) and *Rotylenchus*.

Pratylenchus spp.

The lesion nematodes, *Pratylenchus* spp. (**8, 82, 150, 154**), feed primarily on cortical tissues of smaller, nonsuberized roots. They may also infect tubers, peanut pods, and other below-ground organs. Migration through host tissues and feeding activities results in destruction of host cells with formation of necrotic lesions. All life stages, except the J1, are parasitic and can be found in host roots and surrounding soil (**9**). Eggs are deposited singly in infected tissues and the soil. Sexual reproduction is common for some species such as *P. penetrans* but rare for others such as *P. brachyurus*. The optimum temperature for development is 26°C. As an endoparasite, population densities of *Pratylenchus* spp. are typically much greater in plant roots than in the surrounding soil, and this is a common feature of these migratory endoparasites; therefore, it is essential that samples for diagnosis include roots from symptomatic plants. Soil samples alone may fail to indicate the presence of damaging population densities of these nematodes.

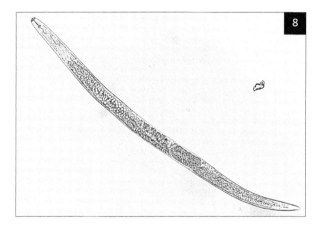

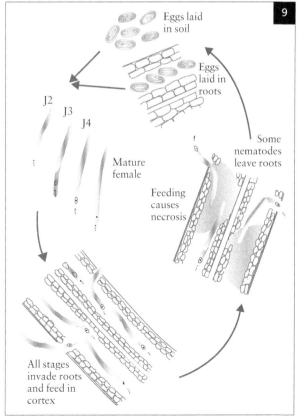

8 *Pratylenchus zeae* female.

9 Life cycle of migratory endoparasites *Radopholus* and *Pratylenchus* spp.

Radopholus spp.

There are several *Radopholus* species damaging crops and they are similar to the lesion nematodes in behaviour and appearance (10, 143). The burrowing nematode, *Radopholus similis* (143) is economically the most important. Burrowing nematodes are aggressive pathogens of such crops as anthurium, citrus (spreading decline disease), banana and plantain (toppling disease), and black pepper (yellows disease). Distinct races (banana race and citrus race) based on host preferences are recognized within this species. The nematode has a relatively large host range, numbering several hundred plant species. Like the lesion nematodes, the burrowing nematode's feeding activities result in the development of necrotic lesions in the infected host tissues (9). All life stages are parasitic and the nematode reproduces sexually. Eggs are laid within infected host tissues (corms, roots, and tubers) and embryonic development is completed in a few days. The entire life cycle can be completed in 3 weeks under optimal conditions. The nematode is restricted to tropical climates.

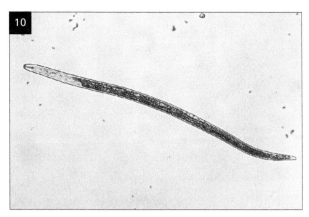

10 *Radopholus citri* female.

Hoplolaimus spp.

The lance nematodes *Hoplolaimus columbus*, *H. pararobustus*, *H. seinhorsti*, and *H. galeatus* (11, 24) are relatively large migratory parasites that feed both endo- and ectoparasitically. All life stages are infective and can be found within the plant roots, feeding on the cortical cells and in the surrounding soil. Feeding activities typically result in cellular necrosis (43). Eggs are deposited singly in roots and the soil. Males of *H. columbus* are very rare in nature, whereas males are common for *H. galeatus*. Diagnosis can be made from either root or soil samples.

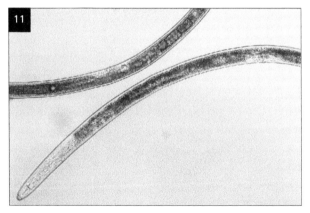

11 *Hoplolaimus seinhorsti* females.

Ditylenchus spp.

Ditylenchus species can have an ectoparasitic phase but are largely migratory endoparasites that infect the bulbs (tulips, narcissus, onion, and garlic), stems (lucerne and clover species), inflorescences (rice), and pods and seeds (peanut) of host plants. All life stages (J2 through to adults) are infective, but the J4 stage is the most common primary inoculum as this stage has the greatest potential for survival of adverse conditions. The nematodes feed primarily on parenchyma tissues, secreting large amounts of cell wall degrading enzymes that cause much tissue destruction. Although many *Ditylenchus* species are able to utilize a variety of fungi as hosts, one of the most important plant pest species of the genus, *D. dipsaci*, appears to be a strict obligate parasite of plants. The host range of *D. dipsaci* includes such diverse crops as oat, lucerne (alfalfa), clovers, tulip, narcissus, onion, garlic, broadbean, strawberry, and many weed species. Many populations of *D. dipsaci* exhibit distinct host preferences. A 'giant race' exists that is particularly important in the Mediterranean region and is distinguished by its larger size. Reproduction is by cross-fertilization and males are common. Eggs are deposited singly in infected tissues. In the later stages of disease, the nematodes aggregate into a mass often referred to as 'nematode wool' (**41**). With slow drying, the J4 stage can enter an anhydrobiotic state and can survive for years in this condition. The survival potential and large host range enables long-term persistence of populations in infested fields.

All stages of *D. dipsaci* are vermiform. The female is 1 mm in length with a posterior ovary and a tapering, pointed tail. It has a flattened head and a small, delicate stylet about 10 µm long. The male is similar to the female, apart from sexual structures.

D. angustus is a tropical species and an important pest of rice in south and southeast Asia causing ufra disease (see Chapter 5). It feeds ectoparasitically on stems but also feeds within the inflorescence. *D. africanus* is an economically important pest of peanuts in southern Africa, feeding on underground growing pegs, pods, and seeds.

Sedentary endoparasites

Meloidogyne spp.

Root knot nematodes (*Meloidogyne* spp.) are all sedentary endoparasites. The wide host ranges of the four most common species (*M. arenaria*, *M. hapla*, *M. incognita*, and *M. javanica*) combined includes nearly all crop plants. The total number of described species is approaching 100, thus many of the populations formerly believed to be one of these four common species are now recognized as a distinct new species. Many species have more restricted or less well described host ranges. Only the second juvenile stage (J2) (**158**) is infective, with root tips being the primary infection court (**12**). The J2 migrate through cortical tissues and establish a permanent feeding site in the region of differentiation of vascular tissues. In addition to roots, the nematodes may also infect other below-ground tissues such as potato tubers, peanut pegs and pods, and rhizomes, corms, or bulbs of various crops. The feeding site of each nematode consists of several host cells that are induced by the nematode to become enlarged, specialized feeding cells called giant cells. The giant cells are plant transfer cells with a dense cytoplasm; they are multinucleate and have elevated rates of metabolism. The J2 develops from the typical vermiform-shaped, pre-parasitic stage to a swollen, sausage-shaped advanced J2, and moults to the third stage (J3), fourth stage (J4), and finally the adult female which undergoes a period of rapid growth to achieve the typical rounded, pear shape (**12, 13, 157**). Egg production begins shortly after the final moult, and up to 1000 eggs can be produced by each female and deposited into a gelatinous matrix. This egg mass usually ruptures through the root cortex and is visible on the root surface. Egg masses are initially light brown in colour but become progressively darker with age.

Meloidogyne spp. are called root knot nematodes because of the characteristic galling of host roots at each infection site. The galls are due primarily to hyperplasia and hypertrophy of cortical tissues

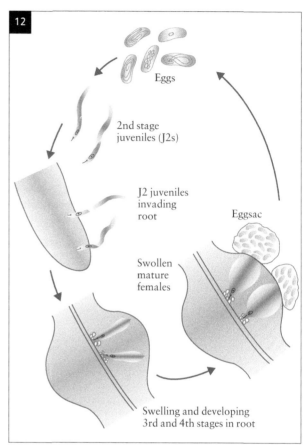

12 Life cycle of the sedentary endoparasite *Meloidogyne* spp.

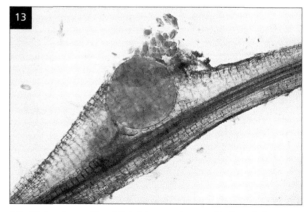

13 Stained, swollen *Meloidogyne incognita* female in root.

surrounding the nematode and the giant cells. Root galling begins within a few days of infection due to a proliferation of cortical cells surrounding the developing nematode. For crop species with succulent roots (e.g. tomato, melons, and cucurbits), small root galls are visible within a few days of infection. Galls may be indistinct early in the season and for crops with fibrous or woody roots such as cotton, groundnut, and cereal grain crops, but on other crops they can be very large especially in the latter part of the growing season. Galls on leguminous crops can be distinguished from *Rhizobium* nodules, in that the galls are an integral part of the root and cannot be removed without severing the root. *Rhizobium* nodules project from the root surface and can be easily removed without severing the roots. Nodules can be infected by the nematodes, with galls and nodules being indistinguishable.

Under favourable conditions most *Meloidogyne* spp. can complete their life cycles in 4–5 weeks, with the production of several hundred eggs by each female; some species such as *M. graminicola* have a shorter life cycle of less than 3 weeks. Soil population densities can increase more than 100-fold during the growing season. The optimal temperature for development of *M. arenaria* and *M. javanica* is 28°C, whereas the optimum for *M. hapla* is 24°C. Additionally, most populations of *M. hapla* can survive at least brief periods of soil temperatures of less than 0°C, whereas *M. arenaria* and *M. javanica* do not survive freezing temperatures.

Because most stages of development of *Meloidogyne* spp. (juvenile J3 and J4 stages and the adult females) are present only in host tissues, the J2 and adult males are the only stages of the nematode present in the soil. Beginning about 6 weeks after planting, egg masses containing hundreds of eggs each can be easily observed on roots with a hand lens or a dissecting microscope. Juveniles can be extracted from the roots, and vermiform adult males can be extracted from soil or infected roots late in the growing season when population densities are high. Rounded to pear-shaped adult females can be dissected from infected roots. Correct identification of the *Meloidogyne* species is critical for effective management based on crop rotation or host resistance.

Nacobbus spp.

The false root knot nematodes, *Nacobbus* spp., so-called because the root galls caused by these nematodes are very similar to those caused by *Meloidogyne* spp., have a sedentary and swollen, mature female stage (**14, 102**). Other stages, including the immature female stage, behave as migratory endoparasites moving between soil and roots (**101**). Therefore, all stages can be found in the root, and all stages except the mature, swollen female, can be found in the soil. Males are common. The swollen female is elongated rather than round and is often mistaken for a developing *Meloidogyne* female in roots. Eggs are laid in a gelatinous matrix extruded from the root. The nematode is found in cooler areas, for example at high altitudes in the Andes, although can be found as concomitant populations with *Meloidogyne*, making their identification or diagnosis very difficult. Galls are small and rounded, often occurring at regular intervals along the root giving the appearance of rosary beads, and hence the common name of rosary bead nematode or 'rosario'. Several *Nacobbus* species are referred to in the literature and the genus taxonomically is in a state of flux; the species most commonly reported is *N. aberrans*.

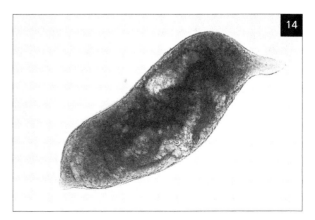

14 Swollen *Nacobbus aberrans* female.

Globodera and *Heterodera* spp.

The cyst nematodes, *Globodera* and *Heterodera* spp., are sedentary endoparasites that have a definite survival stage, the cyst, which is the hardened dead female body containing eggs (**12, 15, 23, 73, 74, 93, 136**). With many of the cyst nematodes, root exudates from host plants are needed to stimulate the eggs to hatch and emerge from the cysts. In *Heterodera* spp., eggs are also laid in a gelatinous egg mass. The second-stage juveniles of both *Globodera* and *Heterodera* (**16A, B, 17**) which emerge from the eggs are the infective stages and invade roots near the root tip, inducing a feeding site of syncytial nurse cells. Nematodes moult through the third and fourth stages and develop to maturity in the root. The females swell and burst through the epidermal layers of the root and become visible on the root surface, first as round, white females and then brown cysts (**17, 22, 33, 73, 135**). No root galling occurs. Males and J2 can be found in the soil, as can the cysts when they eventually break away from the root. Cysts can

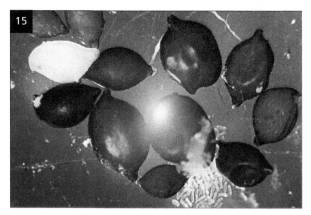

15 *Heterodera sacchari:* brown cysts and white female.

be extracted from the soil by flotation. It is difficult to identify the species, or even the genus, without examination of the mature cysts. The potato cyst nematodes, *G. rostochiensis* and *G. pallida*, are found in cool temperate climates and generally only have one life cycle per growing season. Most of the *Heterodera* species are found in the warm temperate to tropical regions and have a number of generations per cropping season.

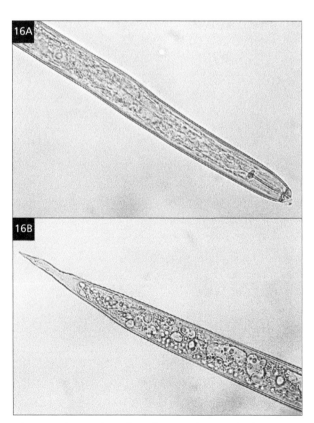

16 Second-stage juveniles of *Heterodera sacchari*.
A: Anterior body; **B:** tail and posterior body.

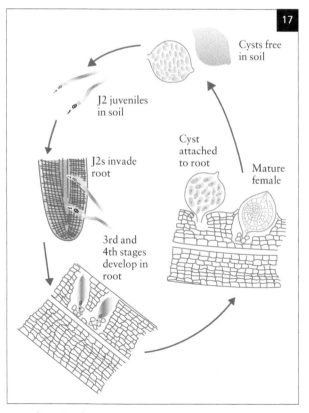

Cysts free
in soil

J2 juveniles
in soil

J2s invade
root

Cyst
attached
to root

Mature
female

3rd and
4th stages
develop in
root

17 Life cycle of cyst nematodes *Globodera* and *Heterodera* spp.

Rotylenchulus spp.

The reniform nematodes, *Rotylenchulus* spp., are sedentary semi-endoparasites. The life cycle of reniform nematodes is unique in that the nematode proceeds from the freshly hatched J2 stage through four moults to the immature female in the soil without feeding (**18**). This development occurs over 7–10 days at temperatures of 25–30°C. The vermiform, immature female is the only infective stage, penetrating the cortex of roots of less than 2 mm diameter with the anterior third of its body. The head of the nematode is then adjacent to the vascular tissue, where the nematode induces formation of multinucleate syncytia that are the parasite's permanent feeding site. These syncytia are functionally similar to the giant cells induced by *Meloidogyne* spp. The posterior two-thirds of the nematode's body swells outside the root (**19**) and assumes the characteristic 'renal' shape within a few days of root penetration (**184**). Reproduction occurs following fertilization of the developing female by males, which are vermiform and apparently develop without feeding. Eggs are deposited into a gelatinous matrix which surrounds the female's body (**185**), with each egg mass containing 50–100 eggs. The nematode can complete several generations in a growing season and population densities at crop maturity often exceed 10,000 nematodes/500 cm^3 soil. Both soil and root samples should be examined for diagnosis. All life stages, except mature swollen females, can be extracted from infested soil. *R. reniformis* has a host range of hundreds of plant species. These nematodes are also able to survive for several months in very dry soils by a process of anhydrobiosis, and the bodies of these dormant forms typically assume tightly coiled positions in the dry soil.

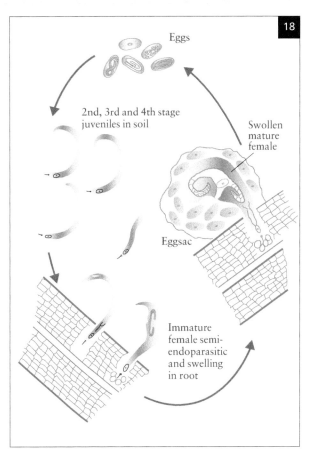

18 Life cycle of the sedentary, semi-endoparasite *Rotylenchulus* spp.

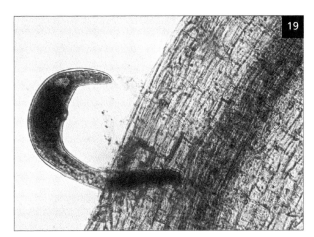

19 *Rotylenchulus reniformis* female, semi-endoparasitic on root and beginning to swell.

Tylenchulus sp.

The citrus nematode, *Tylenchulus semipenetrans*, is also a sedentary semi-endoparasite, almost exclusively of citrus and related plants. The J2 partially penetrate the root and develop through the different stages to mature females, with the neck of the nematode deep inside the root, and a series of feeding or nurse cells are initiated around the head of the nematode. The remaining body of the nematode swells into a characteristic asymmetrical shape on the surface of the root (**20**, **165**). Eggs are laid in a gelatinous matrix around female bodies (**166**) and can cover parts of the root in a continuous layer. These matrices have soil particles adhering to them which obscures the nematodes on the roots. The nematodes are extremely small and are only normally visible when viewed using a microscope and when stained in a suitable stain (see Chapter 8). Only J2 and males can be found in the soil, the juveniles often in very large populations; males are very difficult to identify.

20 *Tylenchulus semipenetrans* females protruding from citrus root.

CHAPTER 2

Grain Legumes

- INTRODUCTION
- SOYBEAN (*Glycine max*)
  Heterodera glycines; *Meloidogyne incognita*,
  M. arenaria, *M. javanica*, and *M. hapla*;
  Pratylenchus agilis, *P. alleni*, *P. brachyurus*,
  P. penetrans, and *P. sefaensis*; *Rotylenchulus
  reniformis*; *Hoplolaimus columbus*;
  Belonolaimus longicaudatus
- PEANUT (GROUNDNUT) (*Arachis
  hypogeae*)
  Meloidogyne arenaria, *M. hapla*, and
  M. javanica; *Pratylenchus brachyurus*
  Godfrey; *Aphelenchoides arachidis* Bos;
  Belonolaimus longicaudatus Rau;
  Aphasmatylenchus straturatus Germani

- OTHER BEANS AND PEAS
  Heterodera goettingiana Liebscher;
  Meloidogyne spp.; *Ditylenchus dipsaci*
  (Kühn) Filipjev; *Hoplolaimus seinhorsti*
  Luc

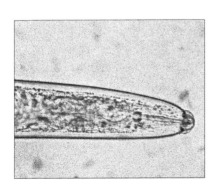
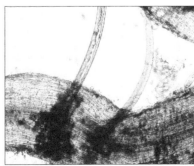

Introduction

There are more than 20 important legume species grown as major food crops in different regions of the world. In addition to the widely grown soybean, *Glycine max*, and peanut or groundnut, *Arachis hypogeae*, there are many other genera and species of legumes commonly referred to as beans or peas that are vital sources of calories, proteins and oils. Nematodes are economically important pests of all the legume crops.

Soybean *Glycine max*

Soybean is an important crop in North and South America, South Africa, and Asia. The most serious nematode parasites of soybean are the cyst nematode, *Heterodera glycines*, and the root knot nematodes, especially *Meloidogyne incognita*, *M. arenaria*, and *M. javanica*. These nematodes are aggressive parasites able to cause significant yield losses and are widely distributed in most areas of soybean production. Nematodes of lesser importance, due primarily to their more restricted distributions, include the reniform nematode, *Rotylenchulus reniformis*, the Columbia Lance nematode, *Hoplolaimus columbus*, several lesion nematodes, *Pratylenchus* spp., and the sting nematode, *Belonolaimus longicaudatus*.

Heterodera glycines

Distribution

The soybean cyst nematode, *H. glycines*, is widely distributed. The first official report of the occurrence of the nematode was from Japan in 1915. It has subsequently been reported from China, Korea, the former Soviet Union, and Taiwan in Asia. In the new world, *H. glycines* is widely distributed in the USA, and is reported from Ontario province in Canada, from Argentina, Brazil, and Columbia. In the USA, the nematode was first reported from a single location in North Carolina in 1954 and is now present from Florida in the south to Minnesota in the north, and from New Jersey in the east to Texas in the west. Most infestations in the new world are believed to have occurred with the importation of soybean seed or soil from Asia (as a source of rhizobia inocula) during the late 19th and early 20th centuries. There is evidence, however, to suggest that the nematode may also be native to North America as a parasite of weed hosts. The nematode occurs across a wide range of temperate to sub-tropical environments and in a wide range of soil types.

Symptoms and diagnosis

Yield suppression of 10–15% without obvious expression of other symptoms is often the first sign that a field is infested with this nematode. Higher levels of infection by the soybean cyst nematode typically cause a distinct chlorosis of the foliage, followed by stunting of the plant (**21**). Affected portions of the field are often first observed as elliptical areas of symptomatic plants. The chlorosis is prominent because the soybean cyst nematode inhibits the development and activity of N_2-fixation nodules. In the most severe infestations, plant stands are reduced. Although the nematode does not cause even discrete root galls, some swelling of the roots at the site of infection may be observed. White swollen females and mature brown cysts can be observed on the root surface, especially with the aid of a hand lens (**22**). The presence of white females and/or mature brown cysts on the roots of soybean is sufficient for diagnosis of the soybean cyst nematode. Cysts, vermiform males and J2 individuals can also be extracted from soil samples throughout the year.

Economic importance

H. glycines is an aggressive parasite that causes yield losses amounting to millions of dollars in the USA annually. The widespread distribution of the soybean cyst nematode contributes to its major economic importance. Losses in the single state of Iowa were estimated at more than 5.8×10^9 kg valued at $13 million in 1997. Similar losses are estimated throughout soybean production areas in the USA and wherever the nematode occurs on the crop. Although damage thresholds vary with soil type, initial population densities of 100 eggs/500 cm^3 soil may suppress yields by as much as 30%. Greatest damage is typically observed in more coarsely textured, sandy soils.

Management

Management of *H. glycines* is achieved primarily though the combined use of resistant soybean cultivars and crop rotations. Several hundred soybean cultivars are available in the USA that have resistance to one or more races of the soybean cyst nematode. However, there are several races for which resistance is not yet available in agronomically acceptable cultivars. The former system of describing races is being discontinued because it was inadequate for describing variation in virulence within a population of *H. glycines*. A new HG Type scheme has been developed that more accurately describes the ability of a given population to parasitize the available sources of resistance. As with other plant

21 Damage to soybean caused by *Heterodera glycines* in the USA. (Courtesy of D.P. Schmitt.)

22 White and light brown coloured cysts of *Heterodera glycines* exposed on infected soybean root. (Courtesy of R.D. Riggs.)

pathogens, continued use of a single source of resistance to *H. glycines* will usually result in a shift in the virulence characteristics of that population, such that that source of resistance is no longer effective. Thus it is critical that use of resistance be limited and used in combination with other management tactics, especially crop rotation. Although *H. glycines* parasitizes numerous weed species, it is only able to parasitize successfully relatively few crop species. Adzuki bean (*Vigna angularis*) and common bean (*Phaseolus vulgaris*) are also good hosts of the nematode in addition to soybean. Therefore, rotation of soybean with a wide range of crops, especially graminaceous, solanaceous, and cruciferous crops, is effective in managing nematode populations. The most effective control is achieved with 2 years of either a nonhost crop and/or an effective resistant soybean cultivar. Good weed control to eliminate potential alternate hosts is an important consideration.

Identification
Second-stage juveniles range in size from 0.375 mm to 0.540 mm, with a prominent stylet (22–25 μm), a sclerotized lip region, a distinct oesophageal intestinal overlap, and an acute tail terminus. The cysts are lemon-shaped with protruding neck and vulva cone, and are of variable size (340–920 μm length × 200–560 μm width) (**23**).

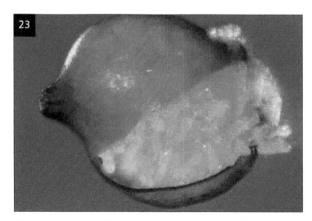

23 Mature cyst of *Heterodera glycines* opened to expose eggs carried in the cyst. (Courtesy of R.D. Riggs.)

Meloidogyne incognita, M. arenaria, M. javanica, and *M. hapla*

Distribution
All four of the root knot nematodes, *Meloidogyne* species, are distributed world-wide in different agroecosystems. In warm temperate and tropical regions, *M. incognita* is the most commonly detected species of root knot nematodes, followed by *M. javanica. M. arenaria* is the least common species. These three species are rarely found where the mean temperature for the coldest month is less than 1°C. *M. hapla* is more common in the cooler temperate zones, and is relatively rare in climates where the mean temperature of the warmest month exceeds 27°C. All *Meloidogyne* species are favored by coarsely textured, sandy soils and are rarely found in finely textured soils with high percentages of silt and clay.

Symptoms and diagnosis
Symptoms caused by all *Meloidogyne* species on soybean are similar, with plants stunted and chlorotic at high levels of infection. Root galling is moderate to severe, varying with level of infection. Generally, galling by *M. arenaria* and *M. javanica* is more prominent than that caused by *M. incognita.* Galls caused by *M. hapla* are small and are characterized by the presence of adventitious roots. Chlorosis of the foliage is less pronounced with *Meloidogyne* species than it is with *H. glycines.*

Root galls are the most diagnostic symptom of these nematodes. Galls can be distinguished from *Rhizobium* root nodules in that nodules are easily broken off the root and may have a pink (active) to greenish (inactive) interior. Nematode galls are an integral portion of the root and cannot be removed without severing the root. Nodules may be galled in later stages of crop development.

Economic importance
The widespread distribution of root knot nematodes is a major reason for their economic importance. *M. javanica* is generally considered to be more aggressive than *M. incognita* which, in turn, causes more damage than *M. hapla.* Aggressiveness of *M. arenaria* varies with host race. Race 1, which also parasitizes peanut, is weakly aggressive on soybean whereas race 2 of *M. arenaria* (that does not

parasitize peanut) is highly aggressive towards soybean. Initial population densities of 150 *M. incognita*/500 cm^3 soil can result in a 10% yield suppression. Greater losses would be expected from *M. javanica* at similar initial population densities, whereas *M. hapla* would cause less yield suppression at this population density.

Management

Management options for root knot nematodes on soybean are dependent upon which species is present, and are made more difficult if the field is infested with multiple species. Several crop rotation systems have been identified for the different *Meloidogyne* species. Cotton is an effective rotation crop with soybean for all species except those populations of *M. incognita* that are parasitic on cotton (host races 3 and 4). Peanut is an effective rotation crop for management of *M. incognita*, most populations of *M. javanica*, and *M. arenaria* race 2. Maize and other cereals are effective in suppressing populations of *M. hapla*. Numerous soybean cultivars are available that have various levels of resistance to *M. incognita*, and a few cultivars have resistance to *M. arenaria* and *M. javanica*. Many of these cultivars also are resistant to one or more races of the soybean cyst nematode.

Identification

Second-stage juveniles are readily extracted from infested soil. They have a slender shape (**158**) with a pointed tail terminus and a delicate stylet (10–12 μm in length for all four species) and the oesophagus has a distinct overlap of the intestine. Mature, pear-shaped females are present only in infected roots and are variable in size and may be as much as 1 mm in diameter (**157**).

Pratylenchus agilis, P. alleni, P. brachyurus, P. penetrans, and *P. sefaensis*

Distribution

The different species of lesion nematodes, *Pratylenchus* spp., are widely distributed especially in coarsely textured, sandy soils. *P. penetrans* is among the most commonly found species in temperate climates, whereas *P. brachyurus* is most common in warmer climates. Other species, including *P. alleni* (north-central USA) and *P. sefaensis* (west Africa) have more restricted distributions, and may be limited to the particular region from which they were initially described.

Symptoms and diagnosis

The primary symptom of damage is the formation of elliptical, necrotic lesions on feeder roots, with a suppression of root and shoot growth with high levels of infection (several hundred nematodes per g of root fresh weight). The amount of necrosis varies with different soybean cultivars and *Pratylenchus* spp. combinations. Diagnosis based on root symptoms is difficult, and usually requires extraction of nematodes from root and soil samples. Collection of root samples is especially important because most of the nematodes are inside the roots during the growth of the crop. Diagnosis based only on soil samples may greatly underestimate the nematode population density.

Economic importance

Lesion nematodes are considered of minor importance to soybean production on a world-wide basis, but can cause substantial yield losses in severely infested fields. Yield losses of 15–20% can be expected in sandy and sandy loam soils.

Management

A few soybean cultivars with moderate levels of resistance to *P. brachyurus* have been identified. Crop rotations can be effective, but require accurate identification of the *Pratylenchus* spp. Maize can be an effective rotation crop for management of *P. brachyurus* but not *P. penetrans*. Cotton is a relatively poor host of both *P. brachyurus* and *P. penetrans*.

Identification

Pratylenchus spp. are relatively short nematodes (0.39–0.80 mm in length), with a short robust stylet (14–22 μm). Species can be distinguished based on variation in the tail shape (conical to crenate or irregular), the number of lip annules, and the position of the vulva (in the posterior third of the body).

Rotylenchulus reniformis

Distribution

The reniform nematode, *Rotylenchulus reniformis*, is widely distributed in tropical and warm temperate regions. These nematodes can be found on soybean in soils of a wide range of textural classes and are commonly found in more finely textured, silty soils.

Symptoms and diagnosis

Reniform nematodes tend to be more uniformly distributed over a field than most nematode species and, hence, the field may lack discrete, irregular areas of symptomatic plants. The first symptom of damage may be a suppression of yield, followed by slight to severe stunting and chlorosis. Reniform nematodes can predispose soybean to seedling diseases when initial nematode population densities are relatively high (greater than 1000 nematodes/500 cm^3 soil at planting). Roots of infected plants may be stunted, but lack other diagnostic symptoms. Signs of the nematode include the presence of immature and mature females protruding from the surface of feeder roots and egg masses covered with soil particles on the root surface (**184, 185** Chapter 7). When heavily infected roots are washed gently with water, the soil particles adhering to egg masses will give the roots a dirty appearance. Soil population densities frequently exceed 10,000 individuals/500 cm^3 soil at crop maturity.

Economic importance

Detectable yield suppression in a range of soil types can be observed when initial population densities of *R. reniformis* exceed 100 nematodes/100 cm^3 soil. Yield suppression of more than 50% has been observed when initial nematode populations exceed 1000 nematodes/100 cm^3 soil.

Management

Management has been primarily through use of granular and fumigant nematicide applications. Soybean cultivars with moderate levels of resistance are available. Although *R. reniformis* has an extensive host range, rotation with nonhosts such as sorghum or peanut provide effective management. Because of the ability of the nematode to survive desiccation, fallowing during a dry season is less effective than having the soil wetted periodically to induce nematode activity.

Identification

Immature, vermiform females of *R. reniformis* extracted from the soil are 0.34–0.42 mm in length, whereas mature, swollen females (reniform = kidney-shaped) from roots are 0.38–0.52 mm in length with a vulva located in the posterior third of the body (**184**). The tail is conical, and the oesophagus overlaps the intestine, typically laterally. Males of *R. reniformis* are 0.38–0.42 mm in length; the bursa envelopes the tail with readily observed spicules.

Hoplolaimus columbus

Distribution
The lance nematode, *H. columbus*, has a limited distribution, being found on soybean primarily in sandy coastal plain soils of the southeastern USA, and has also been reported from Egypt and Pakistan.

Symptoms and diagnosis
H. columbus feeds both as an ectoparasite and an endoparasite on the cortical tissues of soybean roots, causing large necrotic lesions. Affected plants are stunted with reduced root development, and typically occur in irregular clusters within the field. Symptoms are similar to those caused by *Belonolaimus longicaudatus*. Extraction of nematodes from root and soil samples is required for diagnosis.

Economic importance
Yield suppression in the range of 10–38% have been documented in the southeastern USA, when nematode population densities at planting were in the range of 90–200 individuals/100 cm^3 soil. Total yield losses are much lower than for cyst or root knot nematodes because of the limited distribution of *H. columbus*.

Management
Although many common field crops are susceptible to *H. columbus*, rotations with non- or poor hosts such as peanut, sweet potato, tomato, pepper, and other vegetables are effective. Additionally, agronomic practices that reduce other stresses on soybean and promote greater root development (e.g. soil fertility, irrigation, and sub-soiling to disrupt hardpan layers in the soil) are effective in reducing yield losses due to this nematode. Management with granular or fumigant nematicides is rarely justified because of the relatively low profit margins associated with this crop.

Identification
H. columbus is a relatively large robust nematode, with mature females being 1.25–1.6 mm in length, with a heavily sclerotized and prominent lip region. They have a large, robust stylet (40–48 μm) with prominent knobs that resemble a tulip flower (**24**). The tail is blunt and the vulva is positioned near the mid-body.

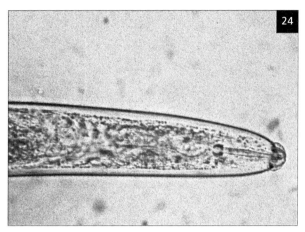

24 Head of *Hoplolaimus columbus* from soybean.

Belonolaimus longicaudatus

Distribution
The sting nematode, *Belonolaimus longicaudatus*, is distributed throughout the southeastern USA, and was recently reported from California on turf grasses, and from the Bahamas, Bermuda, Puerto Rico, Costa Rica, and Mexico. Most surveys indicate that less than 2% of the soybean fields in the southeastern USA are infested with this nematode.

Symptoms and diagnosis
B. longicaudatus induces symptoms similar to those of *Hoplolaimus columbus*, including primarily stunted, chlorotic shoots with poorly developed root systems. Sunken necrotic lesions in the cortical tissues may be visible. Affected plants are typically distributed in irregular clusters throughout the field, with boundaries between healthy and stunted plants usually well marked (25). Seedling death may occur at high initial population densities (greater than 100 nematodes/100 cm^3 soil), resulting in reduced plant populations. Population densities rarely exceed 500 individuals/100 cm^3 soil. Extraction of nematodes from soil is required for diagnosis.

Economic importance
Although *B. longicaudatus* is an aggressive nematode that can cause substantial plant damage where it occurs, it is of overall limited economic importance because of its restricted distribution. As few as 10 nematodes/500 cm^3 soil at planting are sufficient to cause moderate damage to soybean.

Management
Crop rotation with watermelon or tobacco is effective in suppressing population densities of this nematode and increasing soybean yields. Although some populations of *B. longicaudatus* are parasitic on peanut, peanut can be an effective rotation crop in those fields with populations of the nematode that do not reproduce on that crop. Control of *B. longicaudatus* by nematicide application is justified only in the most extreme cases.

Identification
The female *B. longicaudatus* is a relatively long (2–3 mm in length) and slender nematode, with a distinct offset lip region and a long, thin, flexible stylet (100–140 μm) with rounded knobs (6, 132). The vulva is located near mid-body with two outstretched ovaries, and the female tail is comparatively long with a bluntly rounded terminus.

25 Damage to soybean caused by *Belonolaimus longicaudatus* in the USA.

Peanut (groundnut) *Arachis hypogeae*

Peanut (groundnut) is a major food and cash crop in American, African, and Asian countries and has numerous nematode parasites. It is unusual amongst the legumes in producing the mature seed pods and seeds below ground. Among the most common and economically important nematodes of peanut are the root knot nematodes (*Meloidogyne* spp.), the lesion nematode (*Pratylenchus brachyurus*), the testa nematode (*Aphelenchoides arachidis*), the sting nematode (*Belonolaimus longicaudatus*), the pod nematode (*Ditylenchus africanus*), and *Aphasmatylenchus straturatus*.

Meloidogyne arenaria, *M. hapla*, and *M. javanica*

Distribution

The root knot nematodes, *Meloidogyne arenaria*, *M. hapla*, and *M. javanica* are all distributed world-wide. All populations of *M. hapla* are parasitic on peanut and are found parasitizing peanut in the cooler, northern peanut productions areas of Virginia and Oklahoma in the USA, the Punjab state in India, and the Shandong province in China. *M. arenaria* is the most common root knot species attacking peanut in the more southern regions of the USA, in central and southern India, and in southern China. Only race 1 of *M. arenaria* is parasitic on peanut. Most populations of *M. javanica* in the USA are not parasitic on peanut, but some populations that are parasitic on peanut have been reported from a few fields in Georgia and Texas. *M. javanica* populations parasitic on peanut are common in Egypt and India.

Symptoms and diagnosis

All three *Meloidogyne* species cause typical symptoms of nematode damage on the above-ground portions of peanut, including a clustered arrangement of stunted, chlorotic plants with premature senescence (**26**). *M. arenaria* and *M. javanica* cause similar symptoms with respect to root and pod galling. Root galling is indistinct early in the growing season, but becomes more pronounced as the crop matures. Both *M. arenaria* and *M. javanica* can cause severe galling of the pegs and pods, with the pod galls

typically becoming darkly pigmented (**27**). Galling of peanut roots by *M. hapla* is distinguished from that of the other two species in that there is prolific

26 Patches of stunted and chlorotic peanut plants infested with *Meloidogyne javanica* in Egypt.

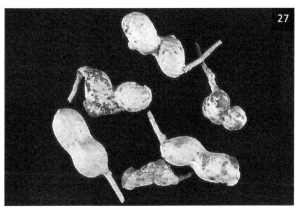

27 Galling and darkening of peanut pods caused by *Meloidogyne javanica*.

adventitious root development from the galls; *M. hapla* rarely forms galls on the pods. Care must be taken to avoid misdiagnosis of *Rhizobium* root nodules as nematode-induced galls. Soil population densities can be near the detection limits during early portions of the growing season. In the middle to the later portion of the growing season, most of the nematodes are developing inside the galled roots or are present as eggs in the egg masses on the root surface. Numbers of J2 stages in the soil are the highest in the later half of the growing season and may exceed 1000 J2/100 cm^3 soil.

Economic importance

Yield losses of more than 50% in heavily infested fields have been documented for *M. arenaria* and *M. javanica*. Losses due to *M. hapla* are usually less and rarely exceed 25% of the yield potential. Several studies have documented a pre-plant damage threshold population density for *M. arenaria* and *M. javanica* of 1–10 juveniles/500 cm^3 soil. The damage threshold population density for *M. hapla* is greater at 16 juveniles/500 cm^3 soil. Because of their widespread distribution and high frequency of occurrence, the root knot nematodes are considered to be very important pathogens of peanut. In some production areas of the USA, nearly 30% of the fields are infested with one or more of these nematodes.

Management

Root knot nematodes on peanut have been typically controlled through the use of nematicides in the USA. The fumigant 1,3-dichloropropene is the most effect nematicide, followed by the nonfumigant nematicides aldicarb and phenamiphos. A common practice is to use either the fumigant 2–3 weeks prior to planting or a nonfumigant at planting, followed by an application of aldicarb about 6 weeks after planting as the pegs are beginning to form. Several crop rotations systems have been shown to be effective for controlling root knot nematodes. Most grass and grain species are poor hosts of *M. hapla* and, therefore, are good rotation crops for this species. Several crops, including cotton, bahiagrass (*Paspalum notatium*), velvet bean (*Mucuna pruriens* var. *utilis*), partridge pea (*Chamaecrista fasciculata*), and American jointvetch (*Aeschynomene americana*),

are effective rotation crops for control of *M. arenaria*. In general, crop rotations are more effective if the non- or poor-host crop is grown for 2 years rather a single year of the alternative crop. No cultivar with resistance to *M. hapla* is available, but the first cultivar (cv 'COAN') with resistance to *M. arenaria* and *M. javanica* was released in the USA in 1999 (28). Biological control of *M. arenaria* with the obligate bacterial parasite *Pasteuria penetrans* has also been effective, but is not yet commercially available.

Identification

The second-stage juvenile (J2) can be identified to genus based on its acute tail, overall length of 0.36–0.56 mm, with *M. hapla* being typically shorter than *M. javanica*, which in turn is shorter than *M. arenaria*. The juveniles are also characterized by having a slender body, a distinct oesophagus overlap of the intestine, and an acute tail terminus. The stylets are delicate with a length of 10–12 μm. The mature females dissected from roots have a distinctly rounded to pear-shaped bodies. The perineal patterns (cuticular markings surrounding the anus and vulva) are helpful in the identification of species.

28 Comparison of the growth of susceptible and resistant peanuts growing in a field infested with *Meloidogyne arenaria*.

Pratylenchus brachyurus Godfrey

Distribution

Lesion nematodes (*Pratylenchus* spp.) are cosmopolitan, and *P. brachyurus* is common in the warm temperate and tropical regions. This species has been specifically reported attacking peanuts in Australia, Egypt, the USA, and Zimbabwe. Additionally, *P. coffeae* has been reported in association with peanut in India.

Symptoms and diagnosis

Diagnostic foliar symptoms are rare, but may include stunting and chlorosis if the level of infection is extremely high. Root symptoms are the presence of distinct necrotic lesions, often elliptical in shape, ranging from a few millimetres to several centimetres in length. Root lesions are most evident on smaller diameter, feeder roots. *P. brachyurus* typically also cause distinct necrotic lesions on the pods (**29**). These lesions are characterized by diffuse rather than sharply delineated margins. Eggs are deposited singly in infected tissues and surrounding soil. The nematode may survive for 24 months in infected pods at room temperature. Diagnosis requires identification of the nematode in addition to observation of symptoms. Because of the endoparasitic nature of these nematodes, detection of the nematode is best accomplished by extraction of root and pod samples.

Economic importance

More than 90% of the total nematode population are typically associated with the host tissues during the cropping season and, immediately after harvest, affecting the yield quantity; *P. brachyurus* also affects marketable value through the effects on pod appearances. Incidences of pod rot caused by soil-borne fungi, especially *Pythium* spp. and *Rhizoctonia solani*, are increased by concomitant infection by the nematode. Aflotoxin contamination of pods due to colonization of nematode-infected pods by *Aspergillus flavus* is also increased.

Management

Losses due to *P. brachyurus* can be reduced in severely infested fields by a 6-week fallow period during the early summer. Timely harvest with removal of infected pods may also reduce nematode population densities

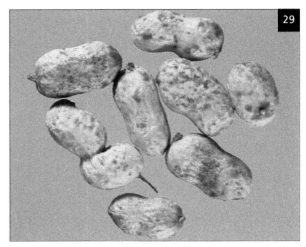

29 Lesions on peanut pods caused by *Pratylenchus brachyurus*.

for the following season. Both fumigant and nonfumigant nematicides are effective in reducing crop losses due to *P. brachyurus*, but are too costly for most peanut production systems. *P. brachyurus* has a wide host range including numerous weed and crop species, making development of effective crop rotation systems difficult. The long-term survival of the nematode in infected pods further complicates development of effective rotation systems. No cultivar with resistance is available.

Identification

P. brachyurus can be identified based on its distinct lip region with two annules, a robust stylet (17–22 μm), and an overall length of 0.75 mm. Adult females, as with all *Pratylenchus*, have a single ovary, with the vulva located posteriorly (80% of body length). The tail terminus is typically blunt. The oesophagus overlaps the intestine ventrally. Males are rarely observed.

Aphelenchoides arachidis Bos

Distribution

The peanut testa nematode, *Aphelenchoides arachidis*, is known currently only from northern Nigeria; however, there is concern that, because it is a seed-borne pathogen, it may be spread to other peanut production areas.

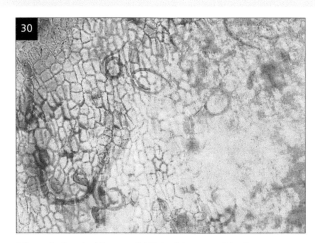

30 *Aphelenchoides arachidis* inside peanut testa.

31 Peanuts infested with *Aphelenchoides arachidis* having brown, shriveled testa compared to healthy confectionery seed.

Symptoms and diagnosis

A. arachidis is parasitic on pods, roots, testa, and other below-ground portions of peanut. Infection results in brown discolouration and necrosis of vascular tissues in the testa (**30**). Mature, dry seeds typically have brown and shriveled testas (**31**). Infected seeds are smaller than noninfected seed, but germination is not affected by the nematode although emergence of seed may be reduced slightly.

The nematode is a migratory endoparasite of peanut, but *in vitro* also feeds ectoparasitically on fungal hyphae. In peanut, *A. arachidis* feeds primarily on parenchymatous tissues, causing cell enlargement and destruction of cell walls. All stages of the nematodes are infectious and eggs are deposited within host tissues. Developing pods may be invaded within 10 days of peg growth into infested soil. The juvenile stages may survive in stored pods for 12 months. Diagnosis is best achieved by extraction of nematodes from symptomatic seed by soaking seeds in water and examination under the microscope.

Economic importance

No data are available on the relationship between initial nematode population densities and pod damage. Primary economic loss arises from reduction in quality of infected seed rather than from suppression of yield quantity.

Management

Destruction of infected shells rather than incorporation into peanut field soil will reduce the likelihood of infection of a new crop from infected seed sources. If hot, dry conditions prevail during harvest, then drying the dug pods in the field prior to storage reduces the severity of the nematode infection. A simple hot water treatment will eradicate nematodes from seeds: seeds are immersed for 5 minutes in four times their volume of water which has been pre-heated to 60°C and removed from the heat.

Identification

A. arachidis has the general characteristics of all *Aphelenchoides* species, a small stylet, large median oesophageal bulb, and rose thorn-shaped spicules in males. It is distinguished from other species by having distinct stylet knobs and a bluntly rounded tail tip.

Belonolaimus longicaudatus Rau

Distribution

The sting nematode *Belonolaimus longicaudatus* is distributed throughout the southeastern USA, was recently reported from California on turf grasses, and has been reported from the Bahamas and Bermuda, Puerto Rico, Costa Rica, and Mexico. Most surveys indicate that less than 2% of agricultural fields in the southeastern USA are infested with this nematode.

Symptoms and diagnosis

Symptoms of parasitism by *B. longicaudatus* include stunted, chlorotic shoots with poorly developed root systems. Sunken necrotic lesions in the cortical tissues may be visible. Affected plants are typically distributed in irregular clusters throughout the field. At high initial population densities (greater than 100 nematodes/100 cm^3 soil), seedling death may occur, resulting in reduced plant populations. Population densities rarely exceed 500 individuals/cm^3 soil. Extraction of nematodes from soil is required for diagnosis.

Economic importance

Although *B. longicaudatus* is an aggressive nematode that can cause substantial plant damage, it is of limited economic importance because of its restricted distribution. As few as 10 nematodes/500 cm^3 soil at planting are sufficient to cause moderate damage to peanut.

Management

B. longicaudatus is most frequently controlled by nematicide application in the USA. Crop rotation with watermelon or tobacco is effective. Cotton, maize or soybean are not effective rotation crops. Summer fallow and flooding will suppress nematode population densities.

Identification

B. longicaudatus is a relatively long (length 2–3 mm) and slender nematode, with a distinct offset lip region and a long, thin, flexible stylet (length 100–140 μm) with rounded knobs (6). The vulva is located near the mid-body with two outstretched ovaries, and the female tail is comparatively long with a bluntly rounded terminus. Males are usually present; the long male tail terminus is acute, with bursa enveloping the tail.

Aphasmatylenchus straturatus Germani

Distribution

This nematode parasite of peanut is known only from the southwestern region of Burkina Faso, Africa. It was initially thought to infest less than 5% of peanut production areas in that county, but this estimate was soon revised to an infestation level of near 25% of the production area.

Symptoms and diagnosis

A. straturatus causes stunting, interveinal chlorosis and reduced rhizobium nodule development. Root systems of severely affected plants are poorly developed.

This nematode feeds both endo- and ectoparasitically, and is motile throughout its life cycle. Nematode population densities are greatest on peanut at 40–70 days after planting, then decline as the peanut matures. Although some of the nematodes feed endoparasitically, typically more than 90% of the population is found in the soil outside the peanut roots. In glasshouse studies, population densities were greater at 30°C than at higher temperatures, and population development is favoured by soil moisture levels of 9–11%. *A. straturatus* is not able to survive anhydrobiotically. Other hosts include cowpea, millet, and sorghum. Extraction of soil samples to confirm the presence of the nematode is needed for diagnosis. Samples from associated wild host karite trees at 40–60 cm depth can be helpful in detection, especially as the peanut crop approaches maturity.

Economic importance

In glasshouse tests, initial population densities of greater than 200/100 cm^3 *A. straturatus* are needed to suppress growth of peanut. In fields studies, initial population densities of 60/100 cm^3 can cause discrete symptoms. Yield losses can be as great as 70% in severely infested fields.

Management

No data are available on successful management methods other than the use of fumigant nematicides. Suppression of nematode populations associated with karite trees may be of aid in management of the nematode on peanut. Glasshouse studies suggest that crop rotations involving cowpea, millet, or sorghum are unlikely to suppress populations of *A. straturatus*.

Identification

A. straturatus belongs to the Hoplolaimidae family of nematodes. It has a strong head and stylet 30–36 μm in length. It is a relatively big nematode around 1.5 mm long, with well defined cuticular markings.

Other beans and peas

In addition to soybean and peanut, there are many other important legume species, commonly known as beans or peas, that are major food crops in different regions of the world. The most commonly grown and significant of these legume crops are the many varieties of the common bean, *Phaseolus vulgaris*, with a world-wide production of 22 million metric tons in 2000. Other species of *Phaseolus* are also important in different parts of the world including the moth bean, *P. aconitifolius*, and the mung bean, *P. aureus*. The common garden pea, *Pisum sativum*, produced yields of 18.3 million metric tons in 2000; chickpea, *Cicer arietinum*, and broad bean, *Vicia faba*, each have a world-wide production of 7.9 million metric tons. Other important legume crops include cowpea (*Vigna unguiculata*), pigeon pea (*Cajanus cajan*), winged bean (*Psophocarpus tetragonolobus*), lentil (*Lens culinaris*), hyacinth bean (*Lablab niger*), and grams, *Vigna* spp. These legumes are susceptible to different nematode species, including several species of cyst nematodes (*Heterodera* spp.), lesion nematodes (*Pratylenchus* spp.), the reniform nematode (*Rotylenchulus reniformis*), root knot nematodes (*Meloidogyne* spp.), false root knot nematode (*Nacobbus aberrans*), and the stem and bulb nematode (*Ditylenchus dipsaci*). A range of other plant parasitic nematodes including *Hoplaimus*, *Hemicyliophora*, *Radopholus*, *Trichodorus*, and *Tylenchorhynchus*, is also found associated with these legumes.

Heterodera goettingiana Liebscher

Distribution

The pea cyst nematode, *Heterodera goettingiana*, is probably the most important cyst nematode species attacking garden pea and other legumes. *H. goettingiana* has a world-wide distribution, being particularly important in Europe and the Mediterranean region. This nematode is rarely found in the USA. Other cyst nematodes parasitic on bean or pea include *H. glycines* (see soybean) on common bean, *H. cajani* on pigeon pea, and *H. ciceri* parasitic to chickpea which is limited in its known distribution to Syria.

Symptoms and diagnosis

Plants severely parasitized by *H. goettingiana* are stunted with a reddish or yellow discolouration (32). As with other cyst nematodes, white females and brown cysts of *H. goettingiana* can be observed on the roots of infected plants (33). Uneven growth can be caused by infection with *Heterodera* spp. (34).

The life cycle of *H. goettingiana* is similar to that of other *Heterodera* species. *H. goettingiana* has a relatively low optimal temperature for development (10–13°C), and typically completes only one or two generations in a growing season in cooler climates. In the Mediterranean region, *H. goettingiana* may complete up to three generations per season. Reproduction is by cross-fertilization and males are abundant. Although some eggs are typically deposited in a gelatinous matrix, under adverse conditions all of the eggs produced may be retained in the cyst.

Economic importance

Population densities of *H. goettingiana* as low as 1 nematode/g soil are sufficient to suppress yield of pea, with total yield loss at initial nematode population densities greater than 75 nematodes/g soil. The wide distribution and low damage threshold for this nematode on pea contribute to its economic importance.

32 Stunted and chlorotic bean plant, *Vicia faba*, heavily infested with *Heterodera goettingiana* in Jordan.

33 White females and cysts of *Heterodera goettingiana* on roots of bean, *Vicia faba*.

34 Uneven growth of pigeon pea in a field infested with *Heterodera cajan*.

Management

H. goettingiana is difficult to control with nematicides due to the nematode's activity in cool soils and the low rate of hatch. Although several legumes are good hosts to *H. goettingiana* (including broad bean, vetch, soybean, and lentils) many other legumes are nonhosts (including several clover species, lupine, chickpea); most grain crops are nonhosts. Crop rotation is the principal management strategy, but requires at least 3 years between susceptible crops to be effective. Rotation lengths of 6 years are often recommended.

Identification

H. goettingiana belongs to the Goettingiana group of *Heterodera* species, and can be distinguished from other *Heterodera* spp. by differences in the morphology of the cysts.

HETERODERA GOETTINGIANA LIEBSCHER

Meloidogyne spp.

The root knot nematodes, *Meloidogyne* spp., are major economically important pests of most of the bean and pea crops throughout the world.

Distribution
In the tropics, the species *M. incognita*, *M. javanica*, and *M. arenaria* are the most common and economically important pests of the crops and occur world-wide. In some temperate regions, *M. artiella* is of prime importance. *M. artiellia* Franklin has been reported from England, Italy, Spain, and Syria and is an important parasite of chickpea.

Symptoms and diagnosis
Most *Meloidogyne* species cause extensive galling on the roots of bean or pea crops (35, 36, 37). Root systems are damaged and reduced in size (38) preventing deep penetration of the roots. Severe damage to the roots results in poor yield, stunted growth, and above-ground leaf chlorosis (39). Wilting is also a common symptom in dry conditions. *M. artiellia* root galls on chickpea differ from those produced by other *Meloidogyne* spp. by being small or indistinct, and the egg masses produced on the root surface can be mistaken for cyst nematodes. Maximum populations occur in light, sandy soils, but *M. artiellia* can be found in soils with as high as 40% clay content.

Economic importance
Substantial yield losses of beans and peas are caused by root knot nematodes. Damage and yield loss is more extensive when plants are under additional stress, such as in drought conditions. Yield losses on chickpea have been observed at initial *M. artiellia* population densities in the range of 10–20 nematodes/100 cm^3 soil.

Management
Nematicides can be effective in suppressing populations of *Meloidogyne* but rarely are warranted economically. Resistance has been found in the common bean, *Phaseolus vulgaris*, and cowpea, *Vigna unguiculata*, to *M. incognita*, and some commercial cutivars are available. Rotations with nonhost crops are effective but some of the root knot species, particularly *M. incognita* and *M. javanica*, have very wide host ranges. Rotating with graminaceous plants and peanut can be effective with these two species but not with *M. arenaria* race 1, which is a pest of peanut. *M. artiellia* has a different host range; chickpea, wheat, and barley are good hosts for this species but crops that are nonhosts include cotton, sugarbeet, oat, maize, tomato, and melon and these can be effective in a rotation system.

Identification
There are differences in the tails of second-stage juveniles which can be used to distinguish the species of *Meloidogyne*, in addition to comparing the perineal patterns on the basal portions of the swollen females.

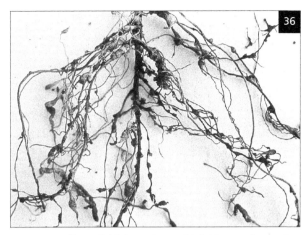

35 Severe galling of cowpea, *Vigna unguiculata*, in Nigeria, caused by *Meloidogyne incognita*.

36 Root galling of chickpea due to *Meloidogyne incognita*. (Courtesy of P. Dart.)

37 Massive root galling of lima bean, *Phaseolus lunatus*, due to infection by *Meloidogyne incognita*.

38 Galled and reduced root system of common bean, *Phaseolus vulgaris*, caused by *Meloidogyne incognita* compared to healthy root, free of the nematodes.

39 Yellowed patches of pea, *Pisum sativum*, caused by *Meloidogyne incognita* in Bolivia. (Courtesy of S.L.J. Page.)

Ditylenchus dipsaci (Kühn) Filipjev

Distribution
The stem and bulb nematode, *Ditylenchus dipsaci*, is mainly confined to temperate regions of the world but can be found in most continents at altitudes in the tropical countries and as a parasite of winter annuals. The nematodes represented by this species are a complex of numerous sub-species and host races. *D. dipsaci* has many forms and a host range of more than 500 plant species world-wide. *D. dipsaci* is a major pest of faba bean (*Vicia faba*) in Europe and the Mediterranean region.

Symptoms and diagnosis
Common symptoms on garden pea and broad bean include swollen, distorted stems and petioles, with distinct lesions that are reddish brown to black in colour (**40**). Lesions on the leaves may be confused with those of various fungal pathogens. Pods and seeds may also exhibit distortion and discolouration (**41**). Plant death occurs in extreme cases. Infection by *D. dipsaci* is often accompanied by infection from secondary pathogens, especially in later stages of disease development. Large numbers of nematodes (hundreds per g tissue) can be extracted from infected plant tissue. Extraction of different life stages from the soil by common extraction protocols is possible, but as there are many other nematodes morphologically similar to *D. dipsaci* in the soil it can be difficult to distinguish species.

Economic importance
Despite the importance and clearly observable damage symptoms on faba bean, few estimates of damage thresholds are available. A damage threshold of 2 nematodes/g soil has been determined for onion. The giant race is considered to be more aggressive on broad bean than is the common 'oat race'.

Management
Use of certified, nematode-free seed is an important and effective means of avoiding disease in fields not already infested with *D. dipsaci*. Despite the large host range, effective crop rotation systems can be developed. Most effective rotation systems require 4 years between plantings of a susceptible crop, along with effective weed control.

Identification
Ditylenchus spp. can be difficult to identify, especially from mixed soil populations, for those without experience working with these nematodes. The bodies are typically attenuated, adults are 1.0–2.2 mm in length, with a thin cuticle. Stylets are delicate, ranging from 10–13 μm for males and females. The females have a single ovary and the vulva is posterior at 76–86% of the body length. The tail is moderate in length, conical, with a rounded terminus. Because of the large number of species within the genus, species identification usually requires assistance from persons with specific training and experience.

Hoplolaimus seinhorsti Luc

Hoplolaimus seinhorsti is one of the many other genera and species of plant parasitic nematodes found on grain legumes which has received relatively little attention but can be locally important as a pest of the crops. In Africa, it has been found as an endoparasite invading and seriously damaging roots of cowpea, *Vigna unguiculata* (**42, 43**).

41 Distorted and darkened seeds of *Vicia faba* infested with *Ditylenchus dipsaci* showing white 'eelworm wool' in pods.

40 Brown and swollen stem of *Vicia faba* due to infection by *Ditylenchus dipsaci*. (Courtesy of D.J. Hunt.)

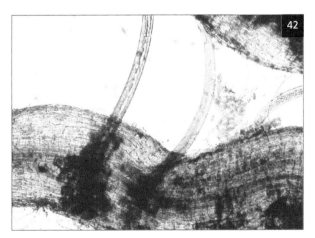

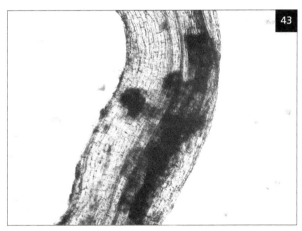

42 *Hoplolaimus seinhorsti* (stained in methyl blue) shown invading roots of cowpea.

43 *Hoplolaimus seinhorsti* endoparasitic in cowpea root surrounded by necrotic tissues.

CHAPTER 3

Vegetables

- VEGETABLE CROPS
  Meloidogyne spp.; *Ditylenchus dipsaci*;
  Rotylenchulus reniformis; *Nacobbus
  aberrans*; Other nematodes of vegetables

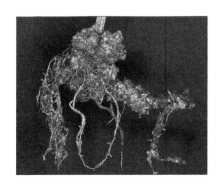

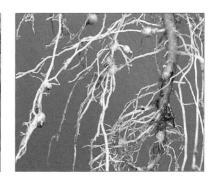

Vegetable crops

Many different types of vegetable crops, fruit, leaf, and bulb, are grown world-wide and have a range of plant parasitic nematodes associated with them. Of particular importance are the root knot nematodes (*Meloidogyne* spp.), followed by others especially the reniform nematode (*Rotylenchulus reniformis*), the false root knot nematode (*Nacobbus* spp.), the stem and bulb nematode (*Ditylenchus dipsaci*), lesion nematodes (*Pratylenchus* spp.), cyst nematodes (*Globodera* and *Heterodera* spp.), stubby root nematodes (*Trichodorus* and *Paratrichodorus* spp.) and sting nematodes (*Belonolaimus* spp.).

Meloidogyne spp.

Distribution

In tropical and warm temperate countries and areas, many of the commonly grown fruit vegetables, such as tomato (*Lycopersicon esculentum*), eggplant (*Solanum melongena*), okra (*Hibiscus sabdariffa*), peppers (*Capsicum* spp.), gourds (*Lagenaria* and *Luffa* spp.), and cucumber and melon (*Cucumis* spp.) are highly susceptible to root knot nematodes, which can be the major pests of the crops. Likewise, other leaf and bulb vegetables such as the brassicas, lettuce, and onions can be severely damaged by different *Meloidogyne* species mainly when grown outside the cool temperate areas, although some *Meloidogyne* species, particularly *M. hapla*, can be important in more temperate areas. The most commonly occurring root knot species on vegetable crops are *M. incognita* and *M. javanica*. A relatively newly described species, *M. mayaguensis*, is also now recognized as an important species on solanaceous vegetables. *M. mayaguensis* is known to occur in the Caribbean basin, Florida in the USA, southern and west Africa, and is found in glasshouse production systems in France.

Symptoms and diagnosis

Root galling on vegetables caused by *Meloidogyne* species tends to be very large and prominent on most of the fruit vegetables (**44–46**) although less so on most chilli peppers (**47, 48**), onions, and leaf vegetables (**49**). Above-ground symptoms are stunted growth, reduced yield, chlorosis (**50**),

44 Severe galling of okra (*Hibiscus sabdariffa*) caused by *Meloidogyne javanica* in Malawi.

45 Large root galls on tomato caused by *Meloidogyne incognita*.

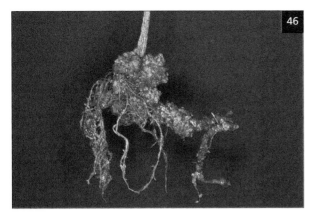

46 Massive galling of snake gourd roots with few normal roots remaining, caused by *Meloidogyne incognita*.

47 Small galls on roots of chilli pepper caused by *Meloidogyne incognita* in Gambia.

48 Relatively small galls covering roots of chilli pepper bush caused by *Meloidogyne incognita* in Bolivia.

49 Small galls covering roots of lettuce caused by *Meloidogyne incognita* in Gambia.

50 Tomatoes in Gambia showing patches of chlorotic and early senescence, symptoms of root knot nematode infestation.

51 Wilting young chilli pepper plants with severe root knot nematode damage in the field in Gambia.

52 Defoliation of chilli pepper bush with severe root galling in Bolivia (see **48**).

53 Eggplant (aubergine) (*Solanum melongena*) with root galls caused by *Meloidogyne incognita* and associated splitting of the fruit in Nigeria.

54 Lettuce with severe root knot galling showing bolting/early senescence symptoms on plants under stress in Tanzania.

wilting (**51**) and defoliation (**52**), all due to the severe root damage. Fruit splitting (**53**) and bolting or early senescence (**50, 54**), can also occur when the plant is under water and nutrient stress because of poorly functioning roots.

On vegetable crops, the *Meloidogyne* species (**13, 157**) generally have the characteristic biology and life cycles of the genus (see Chapter 1). Those occurring in the warmer regions have shorter life cycles, mostly 30–40 days, than the cool temperate nematodes. Uprooting the plants and examining for root galling is a reliable and accurate means of diagnosis. Examining the roots for galls is particularly important when the above-ground symptoms are sometimes identical to other diseases, for example, bacterial wilt which is indistinguishable at first sight from wilt caused by root knot nematodes. It is also important to examine for the slight initial galling that can occur on young seedlings (**55**). Care is needed with the brassicas (cabbage and cauliflower) to distinguish root galling due to the nematodes from club root disease caused by the fungus *Plasmodiophora brassicae*.

55 Young tomato seedlings prior to transplanting in Ghana with initial, slight root galling caused by *Meloidogyne* spp.

Economic importance

The root knot nematodes are major pests of vegetable crops and estimated losses world-wide range from 20–40% on crops such as tomato and eggplant. Local yield losses can be even higher, especially where high soil populations of the nematode occur. If a virulent isolate of *Meloidogyne* is present in the soil, yield losses can be expected if the susceptible vegetable crop is grown too frequently on the same soil.

Management

The most effective means of managing the root knot nematodes are by crop rotation in field soil, producing nematode-free seedlings in the transplanted crops, and the use of resistant cultivars when they exist. The crops that can be used in successful rotations will depend on the species and biological isolate of the *Meloidogyne* present in the soil. Often this can be simply tested in soil from the field by growing different crops and checking for galling. Maize or corn can be a good rotation crop in most situations. Peanut is generally suitable as a crop to follow vegetables when *M. arenaria* is not present in the soil; however, isolates of *M. javanica* that are parasitic on peanut are common in northern Africa and India. Using only nursery beds that are free of root knot will prevent seedlings becoming infected and spreading the nematodes into field soil. Any seedlings with slight galling (**55**) have to be discarded.

Resistance is available in some vegetable crops against root knot nematode species, but not always to all the species that occur on vegetables. Tomato has many resistant cultivars mainly against *M. incognita*, *M. javanica* and *M. arenaria*, but this resistance is not effective against *M. hapla* or *M. mayaguensis*. Resistant cultivars can also be found in carrot, field peas, cowpea, green peppers, eggplant, and okra.

Identification

As can be seen from the discussion above, it can be very important to identify correctly the species of *Meloidogyne* present in the field, although it is not strictly necessary when the crops to be used in rotation are tested in field soil for root galling. Species are distinguished on the basis of the morphology of the infective second-stage juveniles extracted from soil, and on the perineal pattern of the swollen females teased out of the roots.

Ditylenchus dipsaci

The stem and bulb nematode, *Ditylenchus dipsaci*, a nematode of temperate climates including high altitudes in the tropics, is an important pest especially of vegetable bulb crops including onion (*Allium cepa*) and garlic (*Allium sativum*), and to some extent shallots (*Allium ascolonicum*) and leeks (*Allium porrum*). The nematode feeds on both the bulbs and leaves causing deformed growth and the swelling of the bulbs, resulting in a disease condition known as 'bloat'. Leaves also become swollen and twisted with lesions on the surface. In badly infected bulbs, the inner scales become necrotic and brown, seen as circles when the bulb is transversely cut open as occurs in narcissus bulbs (**57–61**, Chapter 4). Bulbs with this amount of infection normally rot in the field or storage. Very large numbers of nematodes can be extracted from around the necrotic areas before the bulbs rot and from the infested leaves, simply by cutting out pieces of tissues and teasing out in a dish of water. Hot water treatment of onion sets to eradicate the nematodes has been successful and the use of nematode-free sets and cloves planted in soil free of the nematode is the most effective means of managing the nematode.

Rotylenchulus reniformis

The reniform nematode has the same distribution and biology on these other vegetables as it does on grain legumes (Chapter 2). It is found in the warmer regions of the southern USA, Africa, Pacific Islands, Middle East, Asia, and south east Asia on all types of vegetables. Its biology on these other vegetables as a sedentary semi-endoparasitic root nematode with a swollen, kidney-shaped female (**18, 19, 184**) is no different to that on other crops. Damage to the roots results in stunted growth and reduced yields.

Nacobbus aberrans

A number of *Nacobbus* species, the false root knot nematodes, has been recognized, but the most commonly identified species on vegetables is *N. aberrans*. *N. aberrans* is indigenous to Latin America and to parts of North America, occurring more in temperate climatic regions such as in the Andes. It has also been found on vegetables, particularly tomato, in glasshouses in Europe. Both the root and above-ground symptoms are similar to that caused by root knot nematodes, but the galls are more discreet and strung along the root (**56**). All stages of *Nacobbus* spp., including the immature female, are infective and can be found in the soil except for the mature, swollen female (**14**) which is sedentary within the cortical tissues of vegetable roots.

Other nematodes of vegetables

The potato cyst nematodes, *Globodera* spp., can be found on tomato and are particularly important where potato and tomato are being grown on the same land. The cabbage cyst nematode, *Heterodera cruciferae*, and the beet cyst nematode, *H. schachtii* (see *H. schachtii* on sugar beet, Chapter 6, **133–136**) can be a problem on cruciferous vegetables in some areas such as the USA. The sting nematodes, *Belonolaimus* spp., the awl nematode, *Dolichodorus heterocephalus*, and the stubby root nematodes, *Trichodorus* and *Paratrichodorus* spp., are also recognized pests of a wide range of vegetable crops in the USA.

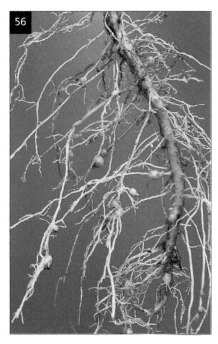

56 Galls of the 'rosary bead' or false root knot nematode, *Nacobbus aberrans*, along roots of tomato.

CHAPTER 4
Flowers

- FLOWER CROPS
  Ditylenchus dipsaci; *Aphelenchoides ritzemabosi*; Other nematodes of flower crops; *Meloidogyne* spp.; *Pratylenchus* and *Radopholus* spp.

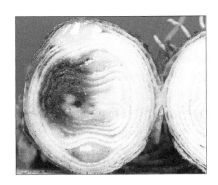

Flower crops

Many flowers are now grown commercially as local or export crops and there is a rapidly increasing trade in cut flowers. Flower crops have a range of associated economically important nematode pests which are mainly bulb and foliar parasites of the genera *Ditylenchus* and *Aphelenchoides*, and root parasites such as *Meloidogyne*, *Radopholus*, *Pratylenchus*, *Heterodera*, *Xiphinema*, *Longidorus*, *Paratrichodorus*, *Criconemoides*, *Belonolaimus*, and *Hoplolaimus* spp.

Ditylenchus dipsaci

The stem and bulb nematode, *Ditylenchus dipsaci*, in addition to being an important pest of *Allium* bulb crops, is also a well known pest of flower bulb crops, especially daffodils, narcissi, and tulips although different biological races of the nematodes occur on the different crops. It is also recorded as a pest of hyacinth, hydrangea, phlox, and penstemon. Another species, *D. destructor*, can be found as a pest of iris, gladioli, and crocus.

Distribution
D. dipsaci is common in Europe and can be found in temperate regions throughout the world where the bulb crops are grown. The nematode is all too easily disseminated in the bulbs used for propagating the crops.

Symptoms and diagnosis
Nematodes feeding in leaf tissues produce very characteristic twisting and malformation of the leaves, and swellings or yellowed blisters (referred to as spickels or spikkels) which can be easily observed and also felt if the fingers are gently run down the leaves (57, 58). Nematodes can be teased out of tissues from around the blisters in a dish of water. They can also be seen *in situ* in stained leaf tissues (59). In the bulbs, feeding produces yellow necrotic lesions initially, which turn dark brown. This necrosis can be observed on the leaf scales within the bulb when cut longitudinally (60) or transversely (61). In advanced stages of disease, the bulbs are also colonized by fungi and bacteria and may be completely decayed.

D. dipsaci is a migratory endoparasitic nematode of leaves, stems, and bulbs (see Chapter 1).

Economic importance
The stem and bulb nematode is a major pest of flower crops in temperate regions and, in all cases where the susceptible flower crops are grown, management of the nematode is necessary.

Management
Chemical, physical, and cultural management methods have all been used to restrict the damage caused by *D. dipsaci*, but chemical treatments are no longer acceptable in most countries where the nematode occurs. In narcissus, hot water treatment of bulbs can be effective in eliminating the nematodes from planting material. Bulbs are pre-soaked at room temperature and then placed in water at 44–46°C for 3 hours. The hot water treatment tank has to have accurate temperature control and good water circulation, as temperatures outside this range will either be inefficient in killing the nematodes or will heat damage the tissues. Hot water cannot be used with tulips. The removal or 'roguing' of infected plants in the field as soon as symptoms appear can reduce levels of infected bulbs. Ensuring that bulbs for planting are from a nematode-free source is the most effective means of preventing losses.

57 Twisted and malformed leaves of narcissus, symptoms of feeding by the stem and bulb nematode, *Ditylenchus dipsaci*.

58 Lesions (spikkels) on malformed leaves of narcissus caused by *Ditylenchus dipsaci*.

59 Stained *Ditylenchus dipsaci* nematodes inside leaf tissues of narcissus.

60 Narcissus bulb cut longitudinally to show lesions and rot of leaf scales caused by *Ditylenchus dipsaci*.

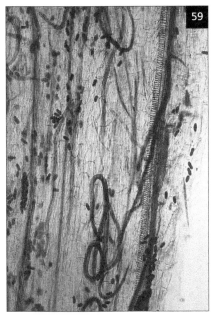

61 Narcissus bulb cut transversely to show brown necrotic lesions in leaf scales caused by *Ditylenchus dipsaci*.

Aphelenchoides ritzemabosi

Distribution

The chrysanthemum leaf nematode, *Aphelenchoides ritzemabosi*, is a pest in temperate countries but has been found at high altitudes in tropical countries such as Africa and South America. It is mainly known as a pest of *Chrysanthemum* species but has also been recorded on many other plants. Three other closely related species, *A. fragariae*, *A. subtenuis*, and *A. blastophorus*, also occur as pests of flower crops. *A. fragariae* produces symptoms similar to *A. ritzemabosi* on begonia, lilies, and violets. *A. subtenuis* has been found on narcissus and *A. blastophorus* causes twisting and distortion of leaves of *Scabiosa* plants.

Symptoms and diagnosis

The nematode is a foliar parasite feeding on leaf mesophyll tissues and growing buds. Cells are destroyed and patches of necrotic tissue spread, at first yellow and then dark brown. The movement of these nematodes is restricted by the veins so each necrotic patch in chrysanthemum leaves is very characteristically angular, initially confined between two leaf veins (**62**), with the veins acting as a barrier to free movement of the nematodes (**63**). Infested leaves eventually die causing a browning of, initially, the lower foliage of the plant (**64**). Although *A. ritzemabosi* is found within tissues, it is not strictly an endoparasite (see Chapter 1). The nematodes survive in the crowns or in dried leaves, but not in the soil.

Economic importance

A. ritzemabosi has been regarded as a major pest of chrysanthemums, especially where the crop is grown under temperate conditions in the field. Infested plants will be reinfected each year from the crowns until there is no flower production.

Management

Modifying the growing conditions for chrysanthemums in controlled cultivation by increasing temperature and using soil-less compound, and nematode-free stocks have greatly reduced the occurrence of the nematodes as a problem. Hot water treatment, 46°C for 5 minutes, can be used to eliminate nematodes from plant crowns.

62 Yellow and brown interveinal necrosis of chrysanthemum leaves caused by the feeding of the leaf nematode, *Aphelenchoides ritzemabosi*.

63 Stained *Aphelenchoides ritzemabosi* nematodes in tissues of chrysanthemum leaf either side of the vein. Note the greater number of nematodes and necrosis in the lower half.

64 Browning of lower leaves of a chrysanthemum plant infested with *Aphelenchoides ritzemabosi*.

Other nematodes of flower crops

Other plant parasitic nematodes known to infest flower crops belong to the genera *Meloidogyne*, *Pratylenchus*, *Radopholus*, *Heterodera*, *Xiphinema*, *Longidorus*, *Paratrichodorus*, *Criconemoides*, *Belonolaimus*, and *Hoplolaimus*. These nematodes are pests in their own right but can also interact with other organisms in disease expression; *Heterodera trifoli* is reported to have a relationship with the wilt causing fungus, *Fusarium oxysporum* f.sp. *dianthi* on *Dianthus* and the strawberry latent ringspot nepovirus on lilies is transmitted by nematodes.

Meloidogyne spp.

Flower crops are hosts for a number of the root knot nematode species. Two species, *M. hapla* and *M. ardenensis*, can be found on flower crops in temperate regions but root knot is mostly a pest of the crops in warmer, tropical regions. *M. incognita* and *M. javanica* are the most frequently recorded root knot nematodes on different flowers both in the field and under glass, including carnations (*Dianthus*), gerbera, *Coleus*, *Heliconia*, *Alpinia*, and *Proteus*. Root galls can be very large in some crops, particularly perennial tree crops such as *Proteus* (**65**).

Pratylenchus and *Radopholus* spp.

These migratory endoparasites cause lesions and rotting of the roots. *Pratylenchus* species occur in both temperate and tropical regions. The main pest in the cooler growing regions is *P. penetrans* which is known to infest many flower crops including *Rosa*, *Phlox*, *Anemone*, *Clematis*, *Delphinium*, *Chrysanthemum*, *Dicentra*, and *Digitalis*. The species *P. vulnus* is a common pest of flowers in warmer regions where many other *Pratylenchus* species are recorded on the crops. *Radopholus similis* is a tropical nematode and can infest roots of flower crops growing in the warmer countries. *R. similis* is of particular importance as a pest of the very susceptible ornamental crop, *Anthurium*.

65 Extremely large root galls on a *Proteus* tree caused by a root knot nematode, *Meloidogyne* spp. in Hawaii.

CHAPTER 5
Cereals

- RICE (*Oryza sativa*)
  Ditylenchus angustus; *Aphelenchoides besseyi*; *Heterodera* spp.; *Hirschmanniella* spp.; *Meloidogyne* spp.; *Pratylenchus* spp.; *Paralongidorus* spp.; *Criconemoides onoensis*

- MAIZE (*Zea mays* L.)
  Pratylenchus spp.; *Paratrichodorus* spp.; *Longidorus breviannulatus*

- WHEAT (*Triticum aestivum*)
  Heterodera avenae; *Meloidogyne* spp.; *Pratylenchus* spp.; *Anguina tritici*; *Ditylenchus dipsaci*

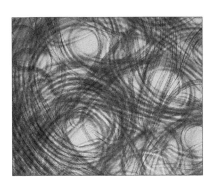 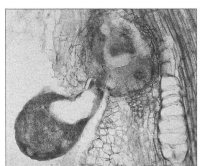 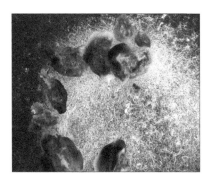

Rice *Oryza sativa*

Rice has a number of important nematode pests that can seriously affect growth and yield of the crop. These nematodes are either foliar or root parasites and there is a wide range of symptoms. Some of the nematodes have a restricted distribution, others occur world-wide. The two foliar parasites are the ufra nematode, *Ditylenchus angustus*, and the white tip nematode, *Aphelenchoides besseyi*. Root parasites are particularly affected by the type of rice being grown; some genera cannot withstand flooding in lowland and deepwater rice and are only found on the upland or dry rice crop. The main nematode root parasites are the cyst nematodes (*Heterodera oryzae*, *H. oryzicola*, *H. sacchari*, *H. elachista*), the rice root nematodes (*Hirschmanniella* spp.), root knot nematodes (mostly *Meloidogyne oryzae* and *M. graminicola*), the root lesion nematodes (*Pratylenchus indicus*, *P. zeae*), the ring nematode (*Criconemoides onoensis*), and the rice needle nematodes (*Paralongidorus australis*, *P. oryzae*). Other species of these and other genera, such as *Tylenchorhynchus*, *Hoplolaimus*, and *Xiphinema* are also reported as parasites on various rice crops.

Ditylenchus angustus

Distribution

Ditylenchus angustus originated as an important pest of deepwater rice in south and southeast Asia and, therefore, has a restricted distribution due to deepwater rice being found in only a few areas. It is a parasite of both wild and cultivated species of *Oryza* in the major river deltas of Bangladesh, Assam, Burma, and Vietnam that are prone to deep or very deep flooding in the rainy season. However, it is now widespread in Bangladesh on lowland rice systems, and may spread into the lowland crop in other countries.

Symptoms and diagnosis

In the vegetative growth stage from seedling to flag leaf, the main symptom of infection by *D. angustus* is leaf chlorosis. Plants become malformed and prominent white patches, or speckles in a splash pattern are seen during vegetative growth at the bases of young leaves (**66**). Necrotic brown stains may develop on leaves and leaf sheaths. Young leaf bases are twisted, leaf sheaths distorted (**67**), and the lower nodes can become swollen with irregular branching. Depending on the severity of infection, chlorotic leaf areas, tillers, or whole plants will wither and die, producing a light brown appearance. When infection is very severe, the whole crop can take on this appearance (**68**). After heading, infected panicles are usually crinkled with empty, shrivelled glumes, especially at their bases; the panicle head and

66 White patches on the growing leaf bases of rice, the initial symptoms of 'ufra' disease caused by *Ditylenchus angustus*, Bangladesh.

67 Twisted and distorted leaves of rice infected with *Ditylenchus angustus*, Bangladesh.

68 Large brown patch of dead rice plants, field symptoms of 'ufra' caused by *Ditylenchus angustus*, Bangladesh.

69 Twisted, distorted, and empty panicles of mature rice due to infection by *Ditylenchus angustus*.

flag leaf are twisted and distorted (**69**). Panicles often remain completely enclosed within a swollen sheath or only partially emerge. Dark brown patches of infected plants can be observed within fields, normally after panicle initiation.

D. angustus occurs in the leaves, inflorescences, young seeds, and rolled stems of growing plants and also in crop residues but it is not seed-borne. The nematode survives in plant residues, the stem tissues, and partially or fully enclosed panicles between seasons in a desiccated state. Nematodes can migrate from diseased plants or plant residues to healthy plants in water, and by stem and leaf contact under high humidity. The nematode is spread primarily through irrigation water, although most nematodes die after a few days in water. *D. angustus* needs at least 75% humidity to migrate on the foliage and is more damaging in wetter areas. Nematodes in water can invade young rice plants within 1 hour. Greatest infection occurs at temperatures of 27–30°C and the nematode has a short life cycle of 10–20 days.

The presence of *D. angustus* can be confirmed by cutting pieces of about 5 mm long from the rolled leaf stems and placing in a small dish of water. Stem pieces are cut longitudinally and left in the water for 24 hours. The rolled leaves or young inflorescence can be teased apart in a Petri dish with water and observed directly. Numerous nematodes (hundreds to thousands) will be active from fresh material, but they require some time to resume activity from dried panicles.

Economic importance

Where *D. angustus* does occur, it can cause total losses in individual fields. In Bangladesh, annual yield losses of 4% have been estimated and in Assam and west Bengal, India, losses are estimated at 10–30% in some areas. It is also recognized as a serious problem in the Mekong delta in Vietnam. Substantial yield losses can occur when transplanted seedlings are infected, even at low percentage infection.

DITYLENCHUS ANGUSTUS

Management

Losses can be minimized by destruction or removal of infested stubble or straw, and burning of crop residues is very effective. Incorporation of crop residues into the soil by ploughing can reduce ufra as nematodes decline quickly in soil. Growing nonhost crops such as jute or mustard in rotation with rice can reduce ufra incidence. Extending the period of time between growing rice crops and lengthening the overwinter period (by delaying sowing, transplanting after flooding, and using early maturing cultivars) can reduce the populations of nematodes and severity of ufra disease on successive rice crops. Removal of volunteer and ratoon rice, wild rice, and other weed hosts will help prevent the carry-over of nematodes to the next crop. Improved water management can help prevent spread of ufra nematodes. Some resistance to *D. angustus* has been found in deepwater rice cultivars and will be effective if commercially available.

Identification

Females of *D. angustus* have thin bodies, 0.8–1.2 mm in length, with fine, pointed tails. The head is light and the stylet is small but distinct. Vulva is posterior. Males are common and are morphologically similar to females.

Aphelenchoides besseyi

Distribution

The nematode is seed-borne and has been disseminated to most rice growing areas of Africa, North, Central, and South America, Asia, eastern Europe, and Pacific Islands.

Symptoms and diagnosis

Characteristic symptoms are the whitening of leaf tips (70). In the field, white patches are apparent from a short distance. Young leaves of infected tillers can be speckled and leaf margins distorted and wrinkled. *A. besseyi* is one of the few seed-borne nematodes. Infected seeds are small and distorted with necrotic lesions.

The presence of *A. besseyi* is detected mainly by soaking the seeds in a dish of water and examining directly using a dissecting microscope (71).

Economic importance

The importance and severity of *A. besseyi* can vary with country, locality, and rice environment. It is considered to be important in India and Africa, but causes little damage in the USA, Thailand, or Japan. In Bangladesh more than 50% of deepwater rice fields have been shown to be infested. Yield loss in susceptible rice depends mainly on the number of nematodes in infected seed and the percentage of nematode infected seed sown; 300 live nematodes/100 seeds has been suggested as an economic damage threshold density.

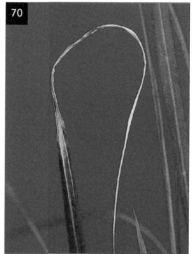

70 White tip of rice leaf caused by *Aphelenchoides besseyi*.

71 *Aphelenchoides besseyi* aggregating in a water drop on a microscope stage. (Courtesy of R. Plowright.)

Management

Prevention by use of clean seed or eliminating nematodes from seed prior to planting is the most effective means of preventing yield loss. Hot water treatment of seed is probably the most useful and cheapest method of reducing crop losses when seed is possibly contaminated. Seeds are pre-soaked in cool water for 18–24 hours followed by immersion in water at 51–53°C for 15 minutes. Higher temperatures (55–61°C) for 10–15 minutes are required if the seed is not pre-soaked. Seed can then be sown directly, or quickly dried and stored.

Identification

The female nematodes are thin, 0.6–0.9 mm in length, with a tapering tail tipped by 3–4 pointed mucron. The lip region is rounded and offset but not strong and the stylet is weak with small knobs (**72**). Vulva is in the posterior part of the body, approximately one-third of the body length from the tail tip.

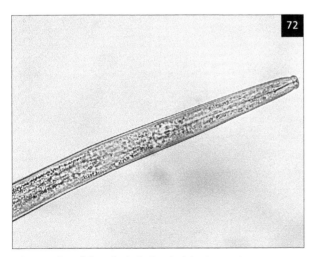

72 Anterior of female *Aphelenchoides besseyi*.

Heterodera spp.

Distribution

Four species of cyst nematodes are known pests of rice in different parts of the world: *Heterodera sacchari* in west Africa and Trinidad, *H. oryzicola* in India, *H. elachista* in Japan, and *H. oryzae* in west Africa and Bangladesh. With the exception of *H. oryzae*, cyst nematodes generally cannot withstand extended flooding and are mainly found on upland rice or on lowland rice where there is little or no water control.

Symptoms and diagnosis

All species cause similar symptoms. Infected plants are severely stunted and chlorotic with less tillers being produced. There is a reduction in root growth, with the roots becoming brown or black. These reduced root systems can appear to have many more small roots stimulated by the nematodes feeding. If soils are heavily infested, seedlings can be killed. Brown cysts and white lemon-shaped females can be observed on the roots without magnification, but the use of a 10× hand lens or a dissecting microscope is helpful.

Economic importance

The cyst nematodes have a limited distribution and are therefore only of local importance, but the cysts can easily be disseminated into new areas. Losses in yield of 20% and 40% have been reported for *H. elachista* and *H. oryzicola*, respectively. Control of *H. sacchari* in the field has resulted in yield increases of over 60%.

Management

Rotation with nonhost plants is an effective means of reducing soil populations. Most crops other than cereals are nonhosts. Resistance to *H. sacchari* has been identified in crosses between *Oryza glaberrima* and *O. sativa*, which could prove useful in the future.

Identification

In all species, the females become obese and lemon shaped, producing brown cysts when dead varying in size from 0.3–1.0 mm in length and 0.2–0.8 mm in width, observable with the naked eye (**15**, **73**, **74**). The active second-stage juvenile is 0.4–0.5 mm in length, with a strong head and stylet and a pointed tail (**16**).

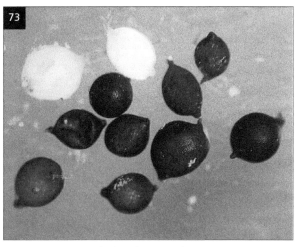

73 Brown cysts and white females of *Heterodera sacchari* from rice.

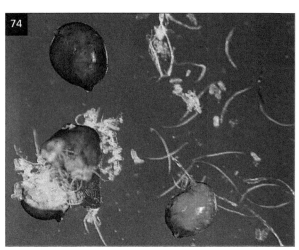

74 Complete and broken cysts of *Heterodera oryzicola* with eggs and second-stage juveniles.

Hirschmanniella spp.

There are many species of *Hirschmanniella* found on a wide variety of crops, and over a dozen of them, known as the rice root nematodes, occur on rice. The most well known species are *H. belli*, *H. gracilis*, *H. imamuri*, *H. mucronata*, *H. oryzae*, and *H. spinicaudata*.

Distribution
Species of *Hirschmanniella* are found in all rice growing areas of the world and comprise one of the few groups of nematodes that can withstand anaerobic conditions such as those occurring in flooded rice fields.

Symptoms and diagnosis
There are no specific foliar symptoms caused by *Hirschmanniella* spp.; root damage produces stunted growth and leaf chlorosis normally seen in clearly defined patches in the field. Tillering and yields are reduced. Nematodes invade roots and migrate through the cortical tissues causing cell necrosis and cavities (**75**, **76**). Infected roots turn brown and rot.

Hirschmanniella spp. are migratory endoparasites feeding on cortical cells of rice roots, invading through the epidermal layers. When roots are examined microscopically, the nematodes, being relatively large, can often be found partially embedded in the tissues in the process of invading (**75**). Eggs are mainly laid

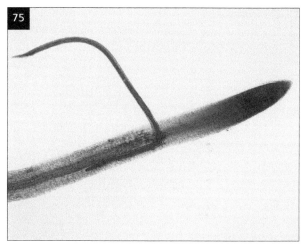

75 Mature female of *Hirschmanniella oryzae* invading rice root.

within the root cortex as the nematodes migrate through the roots feeding on the cells (**76**). Life cycles vary depending on the species, but are generally around 4 weeks from egg to egg.

Economic importance
Because of their ubiquitous occurrence in all rice growing fields, *Hirschmanniella* spp. may cause the most combined yield loss of any of the nematode pests of rice world-wide, although individual field losses may be less. Commonly occurring soil populations of 100–200 nematodes/litre soil can reduce rice yields by 15–30%.

Management
Absence of rice and other susceptible hosts will cause a severe decline in populations. If there is sequential cropping of flooded lowland rice with dry land crops, populations will be reduced but only if the intervals between susceptible crops are more than 1 year. Where green manure legumes are used to increase soil fertility there has been marked control of *Hirschmanniella* spp., especially with *Sesbania*

rostrata and *Sphenoclea zeylanica*, which either act as trap crops or produce toxic root exudates. In lowland rice growing areas where the crop is grown on a regular basis, there are few cultural measures that can be used to manage populations of the nematode. The use of nematicides is not practical or economic.

Identification
Hirschmanniella spp. on rice are long, thin nematodes, females varying from 1 mm to over 3 mm in length. They have strong stylets, generally tapering, pointed tails, and a vulva in mid-body with two outstretched ovaries. Males and females are morphologically similar apart from sexual organs.

Meloidogyne spp.

A range of *Meloidogyne* spp. is parasitic on different types of rice throughout the rice growing areas. The principal species are *M. graminicola*, *M. oryzae*, *M. incognita*, *M. javanica*, *M. arenaria*, and *M. salasi*.

Distribution
M. graminicola causes damage to upland, lowland, deepwater, and irrigated rice and is mainly found in the countries of southeast Asia, but it has also been reported on rice in the USA. *M. oryzae* has only been found in Suriname, South America on lowland irrigated rice. The other species are mainly pests of upland rice and are of less economic importance: *M. incognita*, *M. javanica*, and *M. arenaria* are reported from the Caribbean, Egypt, west and South Africa, South America, and Japan; *M. salasi* only from Costa Rica and Panama.

Symptoms and diagnosis
Symptoms caused by *Meloidogyne* spp. vary according to type of rice and nematode species. Typical symptoms are stunting and chlorosis of young plants combined with retarded maturation, unfilled spikelets, reduced tillering, and poor yield. Symptoms can often appear as patches in the field. All *Meloidogyne* spp., however, cause swellings and galls throughout the root system. Infected root tips become swollen and hooked, which prevents root

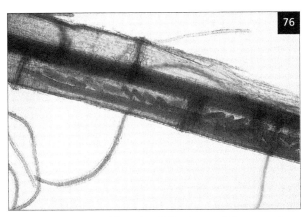

76 Stained eggs and female of *Hirschmanniella oryzae* in the cortex of rice roots.

elongation. This symptom is especially pronounced on rice infected with *M. graminicola* and *M. oryzae* (77). In deepwater rice, *M. graminicola* can cause serious damage prior to flooding and when flooding occurs. Deepwater rice normally elongates rapidly in response to flooding; however, when submerged plants are severely infected with serious root galling they are unable to elongate to the water surface and become drowned out, leaving patches of open water in flooded fields.

The swollen female of *M. graminicola* (78) lays eggs in an egg mass within the cortex of the root (79), unlike other *Meloidogyne* spp. Juveniles hatch from the eggs and reinfect the same root, remaining in the maternal gall or migrating intercellularly through the parenchymatous tissue. This is an adaptation by *M. graminicola*, enabling it to continue multiplying within host tissue under flooded conditions. Juveniles, which migrate from the root in flooded soil, cannot reinvade. The life cycle of *M. graminicola* under optimum conditions at 25–30°C can be as short as 19 days.

M. graminicola can survive in waterlogged soil, initially in rice root remnants, and can remain viable for at least 14 months. The nematode is not active in flooded soils and is unable to invade rice under flooded conditions, but can quickly invade rice plants when soils are drained. *M. incognita*, *M. javanica*, *M. arenaria*, and *M. salasi* are mainly parasites of upland rice and do not survive long periods in flooded soil. *M. oryzae* can survive in shallow flooded (less than 10 cm water depth) rice fields for relatively short periods. *M. graminicola* has a wide host range, which includes many common rice weeds such as *Echinochloa*, *Cyperus*, and *Panicum*. Similarly, *M. incognita*, *M. javanica*, and *M. arenaria* have a very wide host range. A number of weeds and crops are also alternative hosts of *M. oryzae* and *M. salasi*.

Economic importance

M. graminicola is a major pest of all types of rice where it occurs. *M. incognita* and *M. javanica* are of less importance. Damage can be serious on young seedlings raised in well-drained nursery soils

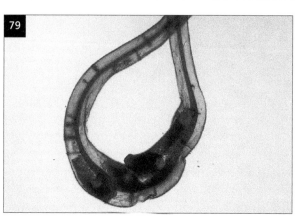

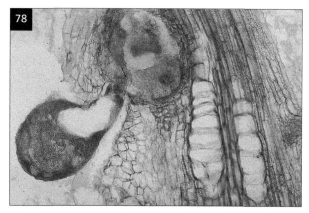

77 Characteristic hooked, root tip galls on rice caused by *Meloidogyne graminicola*.

78 Longitudinal section through rice root showing swollen, sedentary female *Meloidogyne graminicola* feeding on giant cells.

79 Females and egg sacs of *Meloidogyne graminicola* within a developing rice root gall.

before transplanting, and infected transplants will disseminate the nematodes into the field. Destruction of up to 72% of deepwater rice plants by drowning out has been shown to occur with populations of 4000 juveniles/plant of *M. graminicola*.

Management

The type of rice and the species of *Meloidogyne* present will affect the management methods. The use of resistance, controlled flooding, crop rotation, and chemicals can all play a part in the management of the rice root knot nematodes. Although resistance would provide the most sustainable means of control, only relatively few rice lines are truly resistant. Cultivars of the African rice, *O. glaberrima*, are resistant to *Meloidogyne* spp. and progeny being derived from a cross between *O. sativa* and *O. glaberrima* can provide improved acceptable cultivars with resistance to these nematodes. Even relatively short periods of flooding in lowland rice will control *M. incognita*, *M. javanica*, and *M. arenaria* and probably *M. salasi*, but continuous flooding would be necessary for *M. oryzae* and *M. graminicola*. *M. graminicola* will survive normal flooding; however, crop damage can be avoided by producing seedlings in flooded soils. Some crops are reported as poor hosts of *M. graminicola* and, as such, castor, cowpea, sweet potato, peanut, maize, sunflower, sesame, turnip, and okra may be useful in a rotational cropping system. Field application of chemical nematicides is not recommended, but their use in seedling nurseries could possibly prove economically effective.

Identification

The rice root knot species of *Meloidogyne* are distinguished by the posterior cuticular or perineal pattern, and the morphometrics of the second-stage juveniles.

Pratylenchus spp.

Pratylenchus are commonly occurring nematodes and many have been found associated with rice. The most common are *P. zeae* recorded on rice from Africa, North and South America, Australia, and south and southeast Asia, and *P. brachyurus* found on rice in Africa, South America, Pakistan, and the Philippines. *P. indicus* is reported damaging rice in India and Pakistan.

These migratory endoparasites cause lesions and decay of roots resulting in blackened and decreased root systems (80, 81). General nonspecific foliar symptoms are stunting, chlorosis, and reduced

80 Stained *Pratylenchus zeae* nematodes in rice root cortex.

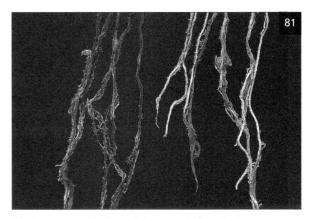

81 Brown and blackened rice roots infested with *Pratylenchus zeae* on the left compared to healthy roots. (Courtesy of R. Plowright.)

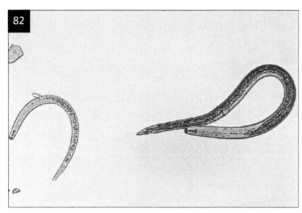

82 Juvenile and female of *Pratylenchus zeae* from rice.

growth and yield. They are parasites of upland rice and only *P. zeae* has been shown to reduce yields. Yields of upland rice have been increased by up to 29% in the Philippines following control of *P. zeae*.

Management can be by fallows and crop rotations, although some *Pratylenchus* species have a wide range of hosts. Reduction of *P. zeae* through crop rotation has been reported using poor- or nonhost crops such as mung bean, black gram, cowpea, and sesame. No resistance has been identified.

P. zeae is vermiform in all stages, the female is 0.5 mm long with a short (15 μm) strong stylet and a posterior vulva (**8, 82, 212**).

Paralongidorus spp.

Paralongidorus are long, thin nematodes, sometimes 10 mm in length. The genus is also characterized by having very long, needle-like stylets. Two species are found on rice, *P. oryzae* on flooded rice in India and Nepal, and *P. australis* in north Queensland, Australia. Symptoms caused by *P. australis* appear 7–10 days after flooding and develop into patches of stunted, yellow plants of which many die. Primary roots show brown necrotic tips, sometimes hooked or curled; secondary roots are shorter than normal, often with a forked appearance and the root system is severely reduced. The life cycle is long, lasting three to four rice crops (about 2 years), and the nematode can survive in flooded, anaerobic soils. Management of the nematodes is

by crop rotation with maize, sorghum, or soybeans; changing from flooded to dry land conditions is a possibility. A decrease in nematode populations can be achieved by increasing the rate of nitrogenous fertilizer in combination with deep ploughing.

Criconemoides onoensis

Studies on *Criconemoides onoensis* in the USA showed that the species can cause severe stunting and yellowing of rice plants. It occurs on both flooded and upland rice. *Criconemoides* species are ectoparasites, feeding on or near the root tip. Parasitized main and secondary roots are stunted and develop characteristic club-shaped tips. This nematode is known to occur on rice in west Africa, Mauritius, Surinam, Belize, and India where it is potentially harmful. *C. onoensis* has a number of alternative weed hosts belonging to the Gramineae and Cyperaceae. *Criconemoides* species are short nematodes, unusual in their external morphology by having prominent annules, hence the common name of ring nematodes (**2**). They also have very long stylets enabling them to penetrate deep into the tissues while remaining external to the root.

Maize *Zea mays* L.

Maize is the third most important grain crop world-wide, following rice and wheat. Maize is grown in most temperate to tropical environments where adequate moisture is available. Numerous nematodes have been associated with maize but, for many, documentation of yield suppression is lacking. Among the most important pests of maize are several species of the lesion nematodes (*Pratylenchus* spp.), the needle nematode (*Longidorus breviannulatus*), and the stubby root nematodes (*Paratrichodorus minor* and *P. porosus*). Other nematodes of lesser importance include the cyst nematodes, *Heterodera zeae* and *H. avenae*, the lance nematode, *Hoplolaimus columbus*, and the sting nematode, *Belonolaimus longicaudatus*. Certain geographical isolates of the root knot nematodes, *M. arenaria*, *M. incognita*, and *M. javanica*, are frequent parasites of maize but cause measurable yield losses only when initial nematode population densities are high (greater than 400 nematodes/100 cm³ soil). However, because these species of root knot nematodes can reproduce well on many maize cultivars, maize production often contributes to yield losses of subsequent crops that are less tolerant of root knot nematodes.

Pratylenchus spp.

Distribution

The lesion nematodes, *Pratylenchus* spp., have a world-wide distribution on maize. *P. hexincisus*, *P. penetrans*, and *P. scribneri* are most common in the more temperate environments such as the central and northern USA, Canada, and Europe, whereas *P. brachyurus* and *P. zeae* are more common in warmer and tropical climates such as the southern USA, Asia, and Africa. All species of lesion nematodes are favoured by coarsely textured, sandy soils. In some regions such as the coastal plains of southeastern USA, nearly 90% of the maize fields are infested with one or more *Pratylenchus* spp.

Symptoms and diagnosis

All species of lesion nematodes are noted for causing necrotic lesions on host roots. Lesions vary in size from less than 1 mm to several centimetres in length, and are often elliptical in shape. Lesion development, however, varies with species. *P. brachyurus* is noted for causing more necrosis of

maize than does *P. zeae*. Foliar symptoms of parasitism by lesion nematodes include stunting and chlorosis where nematode population densities are high (83). Symptomatic plants may occur in irregular clusters within a field, which are often associated with areas of sandier soils.

83 Maize field with damage due to parasitism by *Pratylenchus* spp. in the USA. The centre two rows were not treated, the rows on left and right were treated with the granular nematicide. (Courtesy of W.H. Thames.)

Diagnosis requires identification of the nematode in addition to observation of symptoms. Because of the endoparasitic nature of these nematodes, detection of the nematode is best accomplished by extraction of root samples.

Economic importance
Numerous studies have documented damage to maize due to lesion nematodes, with yield losses in the range of 1200–2300 kg/ha reported for high levels of severe infestation in fields managed for maximum yields. In some instances, a yield suppression of 1% may be observed for each incremental increase of 1000 nematodes/g roots. Loss estimates in less developed agriculture systems with fewer inputs and lower yield potentials are lacking.

Management
Although several nematicides have been shown to increase yields in severely infested fields, such treatments are rarely economically feasible. Resistance to *P. hexincisus* and *P. scribneri* has been identified in some inbred lines, but no hybrids with specific resistance have been developed. Crop rotation can be effective when nonhost crops are available. Cotton is a poor host for many *Pratylenchus* spp. Low tillage cultivation systems have resulted in lower population densities of some lesion nematodes. Fallowing a field can be effective if potential weed hosts are eliminated.

Identification
All the species of *Pratylenchus* on maize are relatively short nematodes (0.4–0.8 mm in length), with a short robust stylet (14–22 mm) and posterior vulva (**82**). Variation in the tail shape (conical to crenate or irregular) can help to distinguish species.

Paratrichodorus spp.

Distribution
The stubby root nematodes, *Paratrichodorus minor* and *P. porosus*, are widespread on maize in the USA, and have been reported from numerous countries in Europe, Africa, several islands in the southern Pacific, India, and the former Soviet Union.

Symptoms and diagnosis
Maize exposed to damaging population densities of these nematodes will have roots that have a distinctly stubby appearance, and can be greatly reduced but without marked discolouration (**84**). The damage to the roots leads to stunting and chlorosis of the shoot growth; however, root symptoms can be mistaken for herbicide damage. Populations reach greatest densities in well drained, coarsely textured soils. As ectoparasites, most nematodes can be found in the rhizosphere soil. Maize is not a host for the viruses transmitted by these nematodes.

84 Healthy maize roots (left) and roots damaged by parasitism by *Paratrichodorus minor* and *Belonolaimus longicaudatus*.

85 Maize field exhibiting damage due to parasitism by *Paratrichodorus minor* and *Belonolaimus longicaudatus*.

Economic importance

Substantial yield suppression can occur when soil population densities of *Paratrichodorus* exceed 150 individuals/100 cm^3 soil (85). Prior to the widespread use of granular insecticides/nematicides on maize, *P. minor* was detected in more than 70% of the maize fields in the sandy soils of the southeastern USA, suggesting that this nematode causes substantial economic losses in that region. In more recent surveys, however, *P. minor* was found in less than 10% of these fields.

Management

Nematicide application for management cannot be justified in most instances due to the relatively low economic value of maize. Rotation with coastal Bermudagrass, garden pea, peanut, soybean, and tobacco can be used to suppress population densities. Because of the sensitivity of the nematodes to desiccation, repeated tillage and a fallow period will reduce population densities.

Identification

Paratrichodorus species can be identified based on the presence of a two-part oesophagus and having a curved stylet. Overall, the body is sausage shaped, i.e. wide relative to length. The body lengths of females are variable (0.50–1.30 mm for *P. minor*).

Longidorus breviannulatus

Distribution

The needle nematode, *Longidorus breviannulatus*, is distributed throughout much of the eastern and central USA. High nematode population densities and yield loss of maize are typically associated with soils with greater than 50% sand content.

Symptoms and diagnosis

In addition to stunting and chlorosis, which may be evident during the first month after crop emergence, affected plants may have a purple discolouration similar to a phosphorous deficiency. Symptomatic plants are usually distributed in oval to elongated patches. Severely affected plants have weakened stalks and reduced ear size. Roots can be yellow in colour with swollen root tips. Feeder root numbers are reduced, giving the root system a coarse appearance. Root symptoms may be confused with symptoms caused by the stubby root nematodes, *Paratrichodorus* spp.

Needle nematodes are noted for having long life cycles, up to 1 year for some species. *L. breviannulatus* achieves maximum population densities at the time of crop maturity, but then may decline significantly during winter months. Because of the large size of these nematodes, they can be difficult to extract from soil. Initial sieving of soil to remove organic debris should be done with sieves having a pore size of 1.65 mm (10 mesh) or 0.8 mm (20 mesh) (see Chapter 8).

Economic importance

Yield suppression can be observed with initial *L. breviannulatus* population densities as low as 10 nematodes/100 cm^3 soil. Seedling death may be observed when population densities exceed 100 nematodes/100 cm^3 soil.

Management

Few specific studies on management of this nematode are available. Nematicide applications are effective but rarely economical. It is likely that alternative nonhost crops can be identified that will permit management by crop rotation. In most crop rotation systems, 2 years of an alternative nonhost are more effective in suppression of nematode population densities than a single season of the nonhost. Clean fallow for several months will also be likely to reduce nematode population densities.

Identification

Longidorus species are readily identified by their large size, with body lengths of the adult females exceeding 2 mm, a two-part oesophagus, an odontostyle lacking basal flanges, and with the vulva near mid-body.

Wheat *Triticum aestivum*

Wheat is one of the three most important grain crops on a world-wide basis. The wheat crop is susceptible to several diseases, arthropod pests, and nematode parasites. The most important nematode parasites are the cereal cyst nematode, *Heterodera avenae*, the cereal root knot nematode, *Meloidogyne naasi*, the lesion nematodes, *Pratylenchus* spp., and the wheat seed gall nematode, *Anguina tritici*. Other cyst nematodes, especially *H. latipons* and *H. filipjevi*, are also damaging pests of wheat in the Middle East and Asia.

Heterodera avenae

Distribution
The cereal cyst nematode, *Heterodera avenae*, is the most important nematode pathogen of cereals in Australia, where it infests nearly 2 million ha. The nematode is also present in Canada, northwestern Europe, India, Japan, the Mediterranean region, and the northwestern part of the USA.

Symptoms and diagnosis
In addition to typical stunting of the shoot growth and foliar chlorosis, a common symptom of the cereal cyst nematode is a stunted, highly branched root system with abundant lateral roots having a thickened appearance. Infection by multiple nematodes at a single point on the roots can result in the roots having a knotted appearance. An important sign of the disease is the presence of young, white, female nematodes and mature dark cysts on the roots. Symptomatic plants are often distributed in distinct patches within areas of apparently healthy plants.

The cereal cyst nematode is adapted to cooler climates with optimum egg hatch occurring at $10°C$, with 40–90% of the eggs hatching. In addition to wheat, other susceptible grain crops are barley (*Hordeum*), oats (*Avena*), and rye (*Secale*). Maize is an intolerant, poor host that supports limited nematode reproduction but may be severely damaged by high initial nematode population densities. Many noncereal members of the grass family Poaceae are also hosts to *H. avenae*. Cysts can be seen with the naked eye and diagnosis is possible by examination of infected roots to confirm the presence of the nematode cysts on the root surface. In addition, diagnosis is aided by observing damage symptoms and by extraction of cysts and second-stage juveniles from infested soil.

Economic importance
Crop damage and yield losses are greatest in Australia and southern Europe where crops are planted at the same time as when greatest egg hatch occurs due to favourable temperatures and rainfall. Damage thresholds vary widely among different regions and with local cereal cultivars, but one study estimated an 87% yield loss of barley with an initial population density of 22 eggs/g soil. Yield losses of greater than 50% have been observed in some wheat fields in Australia and in Asia. Losses in northern Europe and North America are generally less severe.

Management
Management has been achieved primarily through rotations with nonhost crops, especially legumes, and through resistant cultivars. Because of the long-term survival of eggs in cysts, a rotation of at least 2 years is required to maintain population densities below a damage threshold. The effectiveness of resistance varies depending on virulence characteristics of the local population. Approximately 20 nematode biotypes that vary in virulence on different resistance genes in barley, oats, and wheat have been identified. However, many local populations appear to be comprised of only one or a few nematode biotypes. Nematicide applications that result in delayed infection of the roots will limit yield losses.

Nematicides may be most effective when applied only to infested areas of a field that has only recently become infested, thus limiting spread of the nematode. Perversely, continued cropping of infested fields in some instances may increase the population densities of egg parasites of the nematode and result in effective biological control. Unfortunately, it may take as long as 10 years for the natural biological control to suppress nematode population densities to a level that no longer affects crop yield.

Identification

H. avenae is one of more than eight species in the *H. avenae* group. This group of cyst forming nematodes is characterized by lemon-shaped cysts. The second-stage juveniles found in the soil are characterized by a well developed stylet and pointed tail.

Meloidogyne spp.

Distribution

More than 10 *Meloidogyne* species are known parasites of wheat. Of these, the three reported as causing damage are *M. naasi*, *M. artiellia*, and *M. chitwoodi*. *M. naasi* is generally recognized as the most economically important species and has the common name of the cereal root knot nematode. *M. naasi* is widely distributed in Europe and New Zealand, and is present in North and South America; *M. artiellia* is found on cereals in southern Europe and the Mediterranean region, and *M. chitwoodi* in the northwestern portion of the USA, Mexico, Australia, and South Africa. *M. chitwoodi* has been discovered recently in western Europe, but still has a low frequency of occurrence.

Symptoms and diagnosis

Many *Meloidogyne* species, including *M. chitwoodi*, cause only limited or indistinct galls on graminaceous hosts. *M. naasi*, however, causes large terminal, often curved, root galls and spindle-shaped galls on intercalary portions of the roots. Large galls may contain as many as 100 females. Infected root systems may have a bushy appearance and many have limited soil penetration. Severe galling results in stunted plant growth, reduced tillering, and chlorosis of older leaves. Wheat stands can be reduced when initial nematode population densities are high.

M. naasi is unique among the root knot nematodes because of the distinct diapause present in eggs. The nematode typically completes only one generation per season and eggs must be exposed to temperatures less than 10°C for 7 or more weeks before hatch will occur at an optimal temperature of 20–24°C. Additionally, egg masses are often embedded within the gall tissue and not exposed on the root surface. *M. naasi* is more prevalent in humid climates than in semi-arid climates. *M. naasi* reproduces on many graminaceae, but oat is a relatively poor host. Five distinct host races have been described based on reproduction on curly dock (*Rumex crispa*), sorghum (*Sorghum bicolor*), creeping bentgrass (*Agrostis stolonifera*), and common chickweed (*Stellaria media*). Because of the diapause in the eggs, to extract second-stage juveniles from the soil, it may be necessary to store the soil at low temperature (4°C) for several weeks, followed by 5–7 days at room temperature to induce egg hatch.

Economic importance

M. naasi is considered a minor pathogen in many wheat growing regions, but in Italy 17% of durum wheat fields were damaged by this nematode. Continuous cropping of barley in infested fields has resulted in yield declines of greater than 50%, with yields reduced by 3.5% for every increment of 10 juveniles/g soil up to a 50% yield loss.

Management

M. naasi is managed primarily through crop rotations with nonhosts such as potato or sugarbeet or the poor cereal host, oat. No effective resistance has been identified. Fallowing during the spring period conducive for egg hatch is beneficial. Nematicide application is not usually economically feasible.

Identification

As with other *Meloidogyne* spp., species are typically identified by the perineal patterns of the mature females and morphometrics of the second-stage juveniles. *M. naasi* can be distinguished from many other species by the relatively long (0.70 mm) and slender tail of the J2. Assistance from experts in identification of the species is strongly recommended.

Pratylenchus spp.

Distribution
Several *Pratylenchus* spp. are distributed in wheat growing areas world-wide; *P. crenatus*, *P. fallax*, *P. neglectus*, *P. penetrans*, and *P. thornei* are probably the most important on wheat. Damage is most frequently associated with sandy soils.

Symptoms and diagnosis
The most obvious symptoms caused by *Pratylenchus* spp. are the elliptical, necrotic root lesions that vary from 1 mm to several centimetres in length. Infected roots may be slightly swollen. Severe infections limit total root development and lead to stunting and foliar chlorosis. Affected plants in the infested fields may be present in elliptical patches. Tillering may be reduced when initial nematode population densities are very high.

Diagnosis requires identification of the nematode in addition to observation of symptoms. Because of the endoparasitic nature of these nematodes, detection of the nematode is best accomplished by extraction of root samples. More than 90% of the total nematode population are typically associated with the host tissues during the cropping season and immediately following harvest. Population densities of more than 2000 nematodes/g root weight have been observed.

Economic importance
Yield losses are most commonly associated with population densities of greater than 300 nematodes/g fresh root weight. A damage threshold for *P. thornei* has been estimated at population densities at sowing of 50–100 nematodes/100 cm^3 soil. Severe infections of seminal roots may be most damaging. Yield losses rarely exceed 20% in infested fields.

Management
Management of *Pratylenchus* spp. on wheat is difficult because no significant effort has been made to develop resistant cultivars and nematicide applications are not economically feasible. Rotations with nonhost crops is beneficial but requires accurate identification of the *Pratylenchus* species present. Beet and oat are poor hosts for *P. crenatus*, whereas rape controls *P. thornei* and *P. crenatus*. Clean fallow for several months will suppress population densities of all *Pratylenchus* spp., but is seldom a practical solution.

Identification
Pratylenchus species are identified by the lip region, the short robust stylet (15–22 μm), a female body length of 0.50–0.75 mm, the single posterior ovary, and variation in tail terminus shape.

Anguina tritici

Distribution
The ear cockle or seed gall nematode, *Anguina tritici*, is distributed in many parts of the world, but currently has been observed most frequently from north Africa and west Asia. The nematode is mostly found in semi-arid, temperate climates. It was originally an important parasite of wheat in Europe and it has the distinction of being the first plant parasitic nematode ever recognized.

Symptoms and diagnosis
Severe infection of young plants can result in stunted plants with distorted, misshapened stems and leaves (86). The ear or inflorescence may be absent, but when present is wider and shorter. Seeds are transformed into galls that vary in colour from light brown to nearly black (87, 88). Galls caused by the nematode can be confused with symptoms of bunt disease caused by the fungal pathogen *Tilletia tritici*. The galls contain a white powdery mass that consists of dried nematodes in an anhydrobiotic state. Survival of the nematode for up to 40 years in galls has been documented. *A. tritici* is often associated with the spike blight (also called yellow ear rot or tundu disease) caused by the bacterium *Clavibacter michiganense* pv *tritici*.

Diagnosis of ear cockle is based on extraction of nematodes from symptomatic plants, especially apical meristem, and by soaking galls in water which reactivates the many thousands of nematodes found in the gall tissues (89).

86 Twisted and misshapen leaves, early symptoms of damage by *Anguina tritici* on wheat seedling.

88 Healthy seeds and *Anguina tritici* seed galls of wheat.

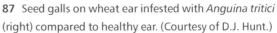

87 Seed galls on wheat ear infested with *Anguina tritici* (right) compared to healthy ear. (Courtesy of D.J. Hunt.)

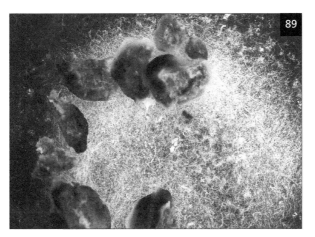

89 Many thousands of *Anguina tritici* nematodes emerging from wheat seed gall soaked in water.

Economic importance

Yield losses of up to 60% of the yield potential of individual fields have been documented, but, in most countries, the nematodes are no longer of economic importance due to the relative ease with which *A. tritici* can now be controlled.

Management

Management of *A. tritici* is readily achieved through the use of seed certification programmes, as the nematode is transmitted in seed contaminated with galls. Because the galls are less dense than healthy seeds, contaminated seed lots may be cleaned by flotation and sieving using fresh or slightly salted water. Rye is also a host but barley, oat, maize, and sorghum are nonhosts.

Identification

The adult females are large nematodes, ranging from 3.0–5.0 mm in length. The body of the female tends to be thickened and curved ventrally (**90**), the stylet is delicate (8–11 µm), and the vulva is located posteriorly (70–95%) with two ovaries.

Ditylenchus dipsaci

The stem nematode, *Ditylenchus dipsaci*, is principally a pest of oats in Europe, USA and Australia, but attacks wheat in central and eastern Europe. It is a migratory endoparasite that can induce increases in cell size and numbers. Symptoms of damage are swollen plants particularly at the base, twisted and distorted stems and leaves, and an increase in number of tillers. Severely infected seedlings can be killed. Damage to the plants is related to climatic conditions and more damage occurs when soils remain moist for extended periods and the temperatures are low. The most economic means of controlling *D. dipsaci* is by growing resistant cultivars.

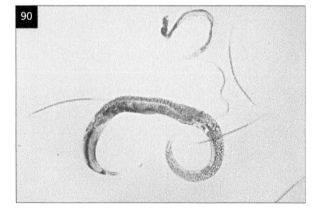

90 *Anguina* female and juvenile nematodes.

CHAPTER 6

Root and Tuber Crops

- INTRODUCTION
- POTATO (*Solanum tuberosum*)
  Globodera pallida, G. rostochiensis;
  Meloidogyne spp.; *Nacobbus aberrans*;
  Ditylenchus dipsaci, D. destructor;
  Lesion nematodes
- SWEET POTATO (*Ipomoea batatas*)
  Meloidogyne incognita; *Rotylenchulus
  reniformis*
- YAMS (*Dioscorea* spp.)
  Scutellonema bradys; *Pratylenchus
  coffeae*; *Meloidogyne* spp.
- CASSAVA (*Manihot esculenta*)
  Meloidogyne spp.; *Pratylenchus
  brachyurus*

- TARO (*Colocasia esculenta*)
  Meloidogyne spp.; *Hirschmanniella
  miticausa*
- GINGER (*Zingiber officinale*)
  Meloidogyne spp.; *Radopholus similis*;
  Pratylenchus coffeae
- CARROT (*Daucus carota*)
- SUGAR BEET(*Beta vulgaris*)
  Heterodera schachtii; *Meloidogyne* spp.;
  Nacobbus spp.; *Ditylenchus dipsaci* ;
  Trichodorus spp., *Paratrichodorus* spp.;
  Longidorus spp.

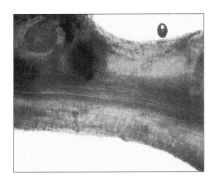
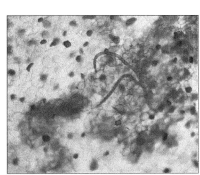
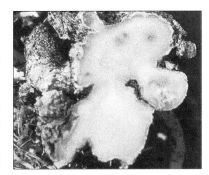

Introduction

Roots and tubers form a very important part of agricultural food production world-wide, and are of particular significance in tropical countries. This is a general description to cover all underground swollen storage organs that can be roots, tubers, rhizomes, or corms. On a world basis, the amount of root and tuber crops grown is second to cereals but, in the tropics, they are the primary source of food carbohydrates. The representative root and tuber crops covered in this chapter are potato, sweet potato, yams, cassava, taro, ginger, carrot, and sugar beet.

Potato *Solanum tuberosum*

The potato tuber is a major food staple for hundreds of millions of people around the world, especially in more temperate climates. Although the potato has its origins in the high altitudes of the Andes of South America, it is now grown in all continents. The most well known, economically important, and possibly the most researched nematode diseases are those caused by the potato cyst nematodes, *Globodera pallida* and *G. rostochiensis*. Several species of root knot and lesion nematodes also cause disease on potato. Other nematode diseases of more limited importance include the false root knot nematode, *Nacobbus aberrans*, and the dry rot disease caused by *Ditylenchus destructor*.

Globodera pallida, G. rostochiensis

Distribution

Because the potato cyst nematodes are parasitic on both tubers and roots and are able to survive adverse environmental conditions, the distribution of these nematodes is nearly equal to the distribution of the crop. *G. pallida* and *G. rostochiensis* originated in the Andes of South America with the potato. These important pests have been reported from 58 countries, representing all continents, in temperate zones and at high altitude in the tropics. Both species are widely distributed in northern Europe, but

G. pallida is more common in southern Europe. Both *G. pallida* and *G. rostochiensis* are present in the USA, but have very limited distribution. Both *G. pallida* and *G. rostochiensis* are present in Canada.

Symptoms and diagnosis

One of the first symptoms of disease caused by the potato cyst nematode is a suppression of tuber yield. With more severe disease due to higher initial nematode population densities, the plants exhibit mild to severe stunting, often occurring in distinct patches within a field. With very high levels of disease, chlorosis and even plant death may occur

(91). The nematodes are visible on the surface of roots and developing tubers, beginning as white, immature females that become darkly coloured cysts with age (92). *G. rostochiensis* goes through a distinct yellow-gold colour as it matures from the white female to the dark brown cyst, giving rise to the name 'golden cyst nematode'; *G. pallida* lacks this distinct golden colour stage.

The life cycle of both *Globodera* species is similar to that of other cyst nematodes. Each female produces 200–500 eggs that are retained completely within the female body and are protected by the transformation of the cuticle of the female to the darkly pigmented, tough cyst (93). The nematode typically completes only one life cycle per crop. Reproduction is inhibited at temperatures greater than 30°C. The eggs within the cyst undergo a distinct diapause, requiring both a period of exposure to temperatures less than 5°C and exposure to potato root diffusates before they will hatch. A cyst may contain viable eggs for more than 10 years; only a portion of the eggs hatch in 1 year, even under optimal conditions. Cysts and second-stage juveniles

are readily extracted from infested soil. Observation of white females and cysts on the roots is diagnostic.

Economic importance

Substantial yield losses are reported from nearly all locations where the potato cyst nematodes are found infecting potato. Losses result from both a reduction in total tuber yield and a reduction in the number of premium size tubers. Potato cultivars can differ widely in tolerance to *G. pallida* and *G. rostochiensis*, but tuber yields can be reduced by more than 20 tons/ha. The damage threshold for the two species appears to be similar and is in the range of 1–2 eggs/g soil. Cultural practices and the environment influence the damage threshold; practices that reduce overall plant stresses tend to increase tolerance to the cyst nematodes.

Management

Management of *G. pallida* and *G. rostochiensis* is achieved primarily through crop rotation and the planting of resistant and/or tolerant cultivars. The host ranges of these two species include several

91 Chlorotic and dying plants due to *Globodera rostochiensis*.

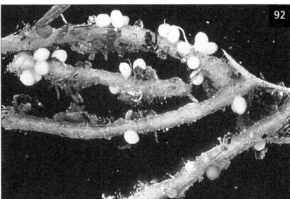

92 *Globodera rostochiensis* on potato roots. (Courtesy of W.F. Mai.)

93 *Globodera* cyst broken to show eggs and second-stage juveniles emerging.

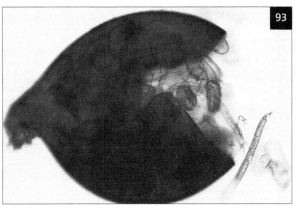

solanaceous weeds and some *Lycopersicon* species; however, most agronomic crops other than potato are nonhosts. Therefore, it is relatively easy to develop cropping systems that suppress the population densities of *G. pallida* and *G. rostochiensis*. Unfortunately, because of the long-term survival capacities of these nematodes, potato can be grown no more often than once every 7 years for rotations to be effective. Effectiveness of rotations is also affected by the presence of weed hosts, especially volunteer potato from a previous crop, during years when the nonhost crops are grown. It is important to control these weeds by cultivation or by application of appropriate herbicides. Resistance is available in numerous potato cultivars, but the effectiveness of resistance is limited by the fact that it is nematode species specific. Accurate identification of the nematode species is critical for effective management. Additionally, multiple races (pathotypes) exist within each *Globodera* species, with the different races attacking different sources of resistance. Nonrace-specific resistance, which is effective against all races, is available in a few cultivars but this resistance provides only partial control. Indiscriminate use of the species- and race-specific resistance can result in shifts in the nematode population such that the resistance is rendered ineffective. Using host resistance and rotation with nonhost crops provides the optimal management.

In Europe, nematicides are widely used as another management tool. The soil fumigants 1,3-dichloropropene and metam sodium are more effective than the nonfumigants aldicarb and oxamyl. The fumigants must be applied 2 weeks prior to planting to avoid phytotoxicity, whereas the nonfumigants can be applied at planting.

Identification

Second-stage juveniles range in size from 375 μm to 540 μm, with a prominent stylet (22–25 μm), sclerotized lip region, and pointed tail. The cysts are round in shape with protruding neck (**93**) and are variable in size (340–920 μm length). There is marked sexual dimorphism, with males being vermiform and 1.0–1.6 mm in length.

Meloidogyne spp.

Distribution

Several *Meloidogyne* species are parasitic on potato with the Columbia root knot, *M. chitwoodi*, and northern root knot, *M. hapla*, being most common in more temperate climates. *M. arenaria*, *M. incognita*, and *M. javanica* can be found attacking potato in more tropical climates. *M. chitwoodi* was first reported from potato in the northwestern USA in 1978 and is mainly confined to that region. It has been reported once from the eastern USA and was introduced into the southwest in 2001 on infected tubers. *M. chitwoodi* has also been found in Argentina, Belgium, Germany, Mexico, South Africa, and the Netherlands.

Symptoms and diagnosis

M. chitwoodi typically produces no or very small galls on potato roots, such that mature females and brown egg masses may be visible on the root surface (**94**). Infected tubers exhibit an uneven surface, having pimple-like galls on the surface (**95**). Internally, the infected tubers may have necrotic spots around the vascular ring, especially just before harvest (**96**). This necrosis may also develop during storage. Similar lesions occur around females of *M. incognita* inside tubers (**97**). *M. hapla* causes more distinct galling, with root proliferation, whereas *M. arenaria*, *M. incognita*, and *M. javanica* cause obvious root galls (**98**) and galled, knobbly tubers (**99**).

M. chitwoodi (temperature optimum 20–25°C) is similar to *M. hapla* (temperature optimum 25–30°C) in that it is adapted to cool temperate climates and survives freezing temperatures better than do *M. arenaria*, *M. incognita*, or *M. javanica*. Although *M. chitwoodi* is active at temperatures less than 10°C, it will not cause galls on tubers until soil temperatures exceed 20°C. A major difference between *M. chitwoodi* and *M. hapla* is in their host ranges; many cereal crops are hosts to *M. chitwoodi*, whereas *M. hapla* is unable to reproduce on most cereal crops and grasses. Root and tuber galling are important diagnostic characters. The adult females can be easily dissected from symptomatic tissues or stained to permit observation of nematodes in the roots (see

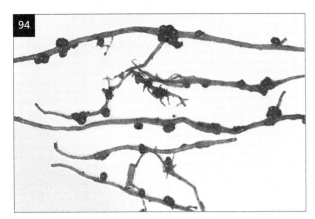

94 Small galls on roots of potato produced by *Meloidogyne chitwoodi*, showing brown egg masses. (Courtesy of J. Clark.)

95 Healthy, smooth potato tuber (**left**) compared to the pimply galls on the surface of the potato tubers caused by *Meloidogyne chitwoodi*. (Courtesy of J.H. Wilson.)

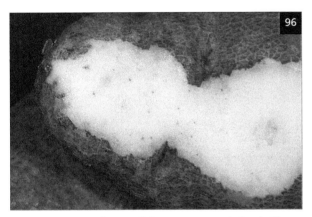

96 *Meloidogyne chitwoodi* damage: external pimple-like lumps on the potato tuber surface, and internal, brown spots. Each spot represents a female nematode with an eggsac. (Courtesy of J.H. Wilson.)

97 Lesions surrounding females and eggs of *Meloidogyne incognita* inside a potato tuber, Bolivia.

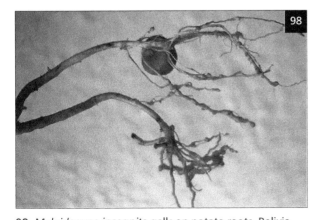

98 *Meloidogyne incognita* galls on potato roots, Bolivia.

99 Galled, knobbly tubers infested with *Meloidogyne incognita*.

Chapter 8). Second-stage juveniles are readily extracted from infested soil.

Economic importance

In temperate climates, *M. chitwoodi* is a much more aggressive nematode than is *M. hapla*, with a reported damage threshold of 2 nematodes/500 cm³ soil, whereas the damage threshold for *M. hapla* is 100 individuals/500 cm³ soil. Because both nematodes infect the tubers, there can be considerable economic loss due to lower quality of the tubers even though yield in tons per hectare may be unaffected at low initial nematode population densities. When the tubers are grown commercially for processing to fried potato chips (crisps) the feeding sites in the vascular tissue of the tubers become dark spots, rendering the final product unacceptable to consumers. Thus in some cases there is a near zero tolerance for nematode infection of the tubers.

Management

Unlike the situation with the cyst nematodes on potato, no potato cultivars are available with resistance to any of the root knot nematodes including *M. chitwoodi*. There is heavy reliance on nematicides in developed countries, with high inputs of these chemicals into potato production. The most commonly used nematicides are the fumigants 1,3-dichloropropene and metam sodium. Nonfumigant nematicides frequently used on potato include aldicarb and ethoprop. Rotation of potato with small grain crops is effective for management of *M. hapla* but is not effective for the more aggressive *M. chitwoodi* because most grain crops are hosts for the latter species. Use of green manure crops, especially rape and sudan grass, are effective in suppressing nematode populations, but the level of suppression is not sufficient to provide a high level of control. Use of green manure plus the nematicide ethoprop applied at planting provides a level of control that approaches that achieved with the fumigant nematicides. Early harvest may help avoid late season increases in nematode population densities that cause the greatest damage to tubers. Tubers infected with *M. chitwoodi* should not be stored as the nematodes will continue to develop at the common storage temperatures of 7–9°C.

Identification

Second-stage juveniles of *M. chitwoodi* have a slender shape with a pointed tail terminus and a delicate stylet (10–12 µm in length). Mature, pear-shaped females are present only in infected roots and are variable in size and may be as much as 1 mm in diameter. Vermiform adult males are relatively rare.

Nacobbus aberrans

Distribution

The false root knot nematode is known primarily from the Americas. It is reported from Argentina, Bolivia, Chile, Ecuador, and Peru in South America, and from Mexico and the western USA in North America. *N. aberrans* originates in this region of the world occurring primarily in the cooler climates of higher altitudes. *N. aberrans* has also been identified from glasshouses in Europe and reported from the former Soviet Union and from India.

Symptoms and diagnosis

Infection of potato roots by *N. aberrans* initially causes small necrotic lesions that are followed later in the growing season by the formation of small, bead-like root galls (**100**). The galls differ from those induced by true root knot nematodes (*Meloidogyne* spp.); the individual galls induced by *Meloidogyne* species are initiated in the centre of small feeder roots and are somewhat symmetrical, whereas the galls induced by *N. aberrans* are in a distinct lateral position, off-set from the central root axis. *Meloidogyne* galls are generally larger and more likely to coalesce spreading along the root, whereas galls of *N. aberrans* are separate like a string of beads. Occasionally, *Nacobbus* and *Meloidogyne* spp. occur together on the same plants and in these cases the galls appear as typical root knot (see **98**). Juveniles (**101**), immature females, and mature males of *Nacobbus* spp. are readily found in soil. When the immature stage becomes established in the root, the female swells to assume the typical swollen body shape (**102**) and the small galls develop around the feeding site.

Economic importance

The false root knot nematode is considered to be the most important nematode pest of potato in several areas of South America, especially at altitudes ranging from 2000–4000 m. Although no precise estimates of the damage threshold are available, tuber yields can be reduced by nearly 90% in severely infested fields, and yield losses greater than 40% are common. Yield losses are usually accompanied by a reduction in percentage of larger sized tubers.

Management

Use of seed tubers free of contamination by *N. aberrans* is essential to prevent further spread of this nematode into fields where it does not already exist. Treatment of seed tubers with fenamiphos or ethoprop can be used to reduce the incidence of transmission on infected seed pieces. Application of large quantities of chicken manure has increased yield in infested fields but does not suppress nematode population densities. The wide host range of *N. aberrans* makes it difficult to develop effective crop rotations, but rotation with oat or broad bean may increase tuber yields without suppressing nematode population densities. Effective control of weed hosts will improve the efficacy of rotation systems. Resistance to *N. aberrans* has been reported, and there is a need for breeding efforts to develop modern cultivars with higher levels of resistance.

Identification

Juveniles, males, and immature females (*ca* 1 mm long) are vermiform with heavily sclerotized lip regions and a strong stylet (20–23 μm) (**101**) and rounded tail terminus. Mature females are swollen and spindle shaped (**102**), the tail terminus often nipple shaped.

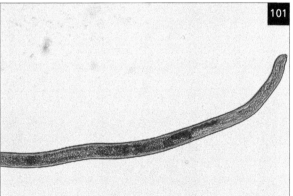

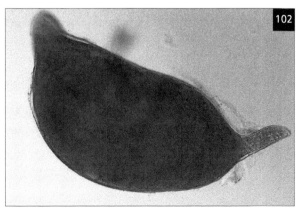

100 Rounded, bead-like galls on potato roots (cv W'aycha) caused by *Nacobbus aberrans*, Bolivia.

101 Vermiform *Nacobbus aberrans* juvenile.

102 Swollen, saccate female of *Nacobbus aberrans*.

NACOBBUS ABERRANS

Ditylenchus dipsaci, D. destructor

Distribution

D. dipsaci has a nearly world-wide distribution, especially in the temperate zones. *D. destructor* is known mostly from Europe, especially eastern Europe.

Symptoms and diagnosis

Both *Ditylenchus* spp. invade the tubers and stolon, with *D. dipsaci* also invading stem and leaf tissues. On the tubers, both species cause a dry rot, with infection by *D. dipsaci* resulting in more superficial symptoms whereas *D. destructor* cause a v-shaped area of dry rot that extends well into the interior of the tuber. Infection of the stem and leaves by *D. dipsaci* may result in stunting and severe distortion of the infected tissues.

Both nematode species may survive in infected tubers, but only *D. dipsaci* is able to survive long periods (months to years) of desiccation by entering the anhydrobiotic state. Large numbers of nematodes can be easily extracted from infected tissues on Baermann funnels or by a similar extraction technique for 24–48 hours (see Chapter 8).

Economic importance

Few data are available on the losses caused by *D. dipsaci* or *D. destructor*, but it is generally accepted that *D. destructor* causes greater losses mainly in eastern Europe.

Management

Management usually begins with the use of seed tubers that are free of infection by these nematodes. Because *D. destructor* does not infect stem tissues, nematode-free tubers for seed can be obtained from stem cuttings that are rooted and planted into noninfested soil. The degree of tuber infection can be reduced by treatment with some nonfumigant nematicides. Rotation to cereal crops for 2–3 years is effective for management of *D. destructor*, but requires that volunteer potato plants in the field be removed to eliminate them as source of carry-over inoculum. Rotation is more difficult with *D. dipsaci* because of its wider host range and the existence of many races that vary in host range.

Identification

Ditylenchus spp. can be difficult to identify. The bodies are typically attenuated, adults are 1.0–2.2 mm for *D. dipsaci* and 0.63–1.9 mm for *D. destructor*. Stylets are delicate, ranging from 10–13 μm for both males and females of *D. dipsaci* and *D. destructor*. The tail is moderate in length, conical, with a rounded terminus. Because of the large number of species within the genus, species identification usually requires assistance from persons with specific training and experience.

Lesion nematodes

Fifteen different species of the migratory endoparasitic lesion nematodes are reported to attack potato roots and tubers, especially in sandy soils. The most common and damaging species are *Pratylenchus penetrans* and *P. scribneri*, which have a nearly world-wide distribution. Symptoms of damage are similar to those caused by this group of nematodes on other crops, with discrete brown to reddish necrotic lesions on the roots (see **200**). Tubers may also be infected, resulting in a roughened coarse surface, with some necrosis of the underlying flesh. Tuber infections are generally restricted to the outer 0.5 mm of tuber. With severe infestations, potato shoots will be stunted and chlorotic. Damage thresholds for *P. penetrans* and *P. scribneri* are approximate 1–2 nematodes/g soil, but is much higher for some species like *P. neglectus*. *P. crenatus* can be present at very high populations densities without any apparent damage to the crop. In addition to direct damage caused by these nematodes, they are known for involvement in root disease complexes with the root rot pathogens *Rhizoctonia solani* and *Pythium* spp. The interaction of several *Pratylenchus* spp. with the vascular wilt pathogen *Verticillium dahliae* results in the serious disease problem called 'early dying syndrome'.

Control of lesion nematodes on potato has been by nematicide applications when the infestations are severe. Effective crop rotations have been difficult to develop due to the wide host ranges of *Pratylenchus* spp. Correct identification of the species present is essential for selection of appropriate rotation crops. Populations of lesion nematodes decline rapidly when fields are fallowed or flooded for several weeks. Although no resistant potato cultivars have been developed, some cultivars have useful levels of tolerance and low levels of resistance.

Sweet potato Ipomoea batatas

Sweet potato is a very widely cultivated carbohydrate food crop of the tropics and sub-tropics, grown for its underground swollen storage roots. The two main groups of nematodes known to be important pests of the crop are *Meloidogyne* spp., especially *M. incognita*, and *Rotylenchulus reniformis*.

Meloidogyne incognita

Distribution
M. incognita is the main root knot species on sweet potatoes and occurs world-wide in the tropics.

Symptoms and diagnosis
M. incognita produces swellings or galling of both normal and storage roots, but generally these swellings often appear only slight compared to galls on other crops (**103**). In cut storage roots, females can be seen surrounded by necrotic cells (**104**). Infected storage roots tend to crack as they mature. Large numbers of nematodes in roots cause poor growth and foliar chlorosis (**105**).

103 Swellings on sweet potato roots caused by *Meloidogyne incognita*.

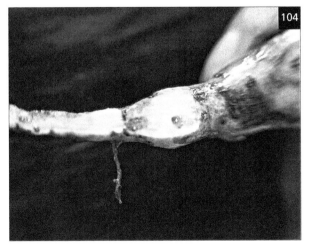

104 Necrosis around *Meloidogyne incognita* females within storage roots of sweet potato.

105 Yellowing of sweet potato in Papua New Guinea due to *Meloidogyne incognita*.

106 Constricted and distorted sweet potato storage root infested with *Meloidogyne incognita*.

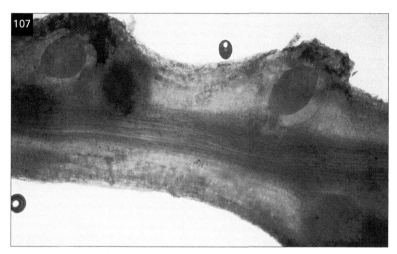

107 Stained *Meloidogyne incognita* females in the cortex of sweet potato roots.

Infested storage roots are constricted and distorted with necrotic patches visible (**106**). Swollen females and egg masses can be found in the galled tissues, often in large numbers strung along the cortex (**107**).

Economic importance
As in other crops, this nematode reduces growth and yield directly by the root damage caused. It also initiates cracking of storage roots and predisposes them to secondary rotting by other pathogens.

Management
Avoiding root knot susceptible crops in rotation with sweet potato reduces build up of the nematodes in the soil. A few cultivars with moderate levels of resistance are available, and growing these cultivars will reduce potential yield losses.

Rotylenchulus reniformis

The reniform nematode is recorded as pest of sweet potato in the USA and can cause cracking of storage roots similar to that caused by root knot nematodes.

ROTYLENCHULUS RENIFORMIS

Yams *Dioscorea* spp.

Many species of yams are grown as carbohydrate food crops, particularly *D. rotundata*, *D. alata*, *D. esculenta*, and *D. cayenensis*. They are vines (**108**) and produce underground tubers that can be stored and therefore are an important source of food during the dry or unproductive season. *Dioscorea* species are also grown as a source of diosgenin which is used in the manufacture of oral contraceptives. Yams are damaged by the nematodes *Scutellonema bradys*, *Pratylenchus coffeae*, and *Meloidogyne* spp. With all these nematode pests, it is mainly the tuber stage of the yam crop that is most seriously damaged.

108 Yam vines, Papua New Guinea.

Scutellonema bradys

Distribution
Scutellonema bradys is commonly referred to as the yam nematode. It is a nematode indigenous to west Africa, which has so far spread to other yam growing areas of the world, mainly in the Caribbean. It has been identified from yams in the African countries of Cameroon, Côte d'Ivoire, Gambia, Ghana, Guinea, Nigeria, and Togo; in the Caribbean in Cuba, Dominican Republic, Guadeloupe, Guatemala, Haiti, Jamaica, Martinique, and Puerto Rico; it has also been reported from Florida, Brazil, India, and Korea. The nematode can also often be found in yam tubers in markets around the world which have been imported from these yam growing areas.

Symptoms and diagnosis
A very characteristic disease of yam tubers known as 'dry rot' is caused by *S. bradys*. This dry rot disease mainly occurs in the outer parenchymatous tissues of tubers extending 1–2 cm into the tuber, although

sometimes deeper, and can be easily seen when the tubers are cut (**109, 110**). The first symptoms of nematodes feeding endoparasitically in the tuber tissues are light yellow lesions immediately below the outer skin of the tuber. These lesions spread around the tuber as the nematodes continue feeding and migrating, and the infected tissues change from yellow to light brown and then turn dark brown to black. In these later stages of dry rot, external cracks appear in the skin of the tubers (**111**) and parts can flake off exposing patches of dark brown, dry rot tissues underneath (**110**). Severe symptoms of dry rot are more readily seen in older and mature tubers. This is especially so during storage (**112**) when dry rot is often associated with other rots and a general decay of tubers (**109**). The nematodes also feed in root tissues but no foliar symptoms have been observed on yams growing in soil infested with *S. bradys*.

S. bradys is a migratory endoparasite, similar to *Radopholus* and *Pratylenchus* spp., of both yam roots and tubers, and can also be found in soils around yam plants. *S. bradys* invades the young, developing tubers through the tissues of the tuber growing point, alongside emerging roots and shoots, through roots, and also through cracks or damaged areas in the tuber skin. Nematodes feed intracellularly in yam tuber tissues resulting in rupture of cell walls, loss of cell contents, and the formation of cavities within the

109 Healthy tuber of white yam (*Dioscorea rotundata*) compared to a tuber with dry rot caused by *Scutellonema bradys*. Note the secondary internal rot of the diseased tuber.

110 Dry rot of water yam (*Dioscorea alata*) tuber caused by *Scutellonema bradys*.

111 External cracking of different species of yam tubers infested with *Scutellonema bradys*.

112 Stored yam tubers with dry rot caused by *Scutellonema bradys* in Cameroon.

tissues. The nematodes are found on the periphery of the root tissues, mainly confined to the outer 1–2 cm of the tuber. *S. bradys* continues to feed and reproduce in stored tubers after harvesting. Populations can increase 9–14-fold in tubers over a 5–6 month storage period. In tubers with partial dry rot, more nematodes are found in the oldest, apical portions, adjacent to the stems.

Economic importance

Although *S. bradys* can cause up to 30% reduction in tuber weights at harvest due to water loss, most economic losses occur in tubers during storage. Dry rot nematode disease is usually followed by bacterial or fungal wet rot of the tuber tissues, and together these can cause anything ranging up to 100% loss of tubers in the yam barns. In yam growing areas where the nematode is prevalent, losses of tubers can average 50%. Generally only parts of the tubers have rot and these tissues are discarded before cooking.

Management

The most successful means of preventing damage by the nematode is by the use of nematode-free propagative material planted in land free, or relatively free, of the nematodes. Yams are propagated from small, whole tubers or tuber pieces and those without nematode symptoms (dry rot, external cracking or flaking of tuber skin) can be selected for planting. Nematode-free seed tubers can also be produced by the 'minisett' method, i.e. using small squares of tuber pieces cut from clean mother tubers. It is also possible to eliminate nematodes from seed tuber material physically by hot water treatment (immersed in water at 50–55°C for up to 40 minutes); however, this treatment requires relatively expensive equipment and careful manipulation to prevent tissue damage. Reducing soil populations of the nematodes can be difficult by nonchemical, cultural methods such as crop rotations, as yams are often grown in inter-cropped systems. Some crops grown prior to yams in a cropping sequence, such as groundnut, chilli pepper, cotton, tobacco, maize, or sorghum, are poor hosts to the nematode and can reduce populations. Some locally traditional practices, particularly incorporating cow dung and other organic manures into planting mounds and coating seed tubers with wood ash, are reported to reduce populations. Chemical control in the field or tubers is not normally an option because of toxicity and cost problems.

Identification

Nematodes are vermiform and the female is around 1 mm in length with a rounded tail, a well-developed stylet, and the vulva in mid-body (**113**). Males are similar to females.

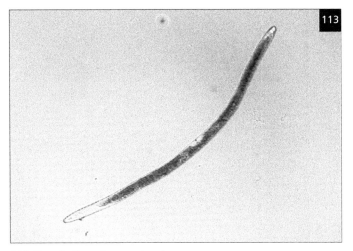

113 Female *Scutellonema bradys*.

Pratylenchus coffeae

Distribution
Pratylenchus coffeae has a wide host range and is a pest on many crops (see banana and plantain, coffee, taro, ginger) and probably originates from the Pacific Island and Pacific Rim countries. The biological isolate that attacks yam has been found on yams in Papua New Guinea, Fiji, Niue, Tonga, Vanuatu and Solomon Islands, in Belize, Puerto Rico, Jamaica, Barbados, Martinique, Guadeloupe, and in China.

Symptoms and diagnosis
P. coffeae causes dry rot symptoms of tubers very similar to those described for *Scutellonema bradys* extending mainly 1–2 cm into the outer tuber, but can penetrate 5 cm into the tuber in some yam species (**114**). Cracking and flaking of tuber skin is common and infected tubers are soft when pressed. Above-ground symptoms are not obvious. Nematodes can be extracted from around dry rot tissues in the tubers to confirm association.

Economic importance
As with *S. bradys*, the nematode is mainly a damaging pest of tubers especially during storage. Incidence in tubers can be high in some regions; dry rot has been found in 30–100% of yam tubers examined in regions of China, in 50% of tubers in Papua New Guinea, and in 67–100% of tubers in the Caribbean. Severe infection and dry rot can prevent sprouting of tubers.

Management
The management methods that are effective for *S. bradys* work also for *P. coffeae*, but cultural control with crop rotations differs because of the different host range of *P. coffeae*.

Identification
P. coffeae is a vermiform nematode, with the sexes morphologically similar and having thin bodies, 0.5–0.7 mm in length. They have flattened heads and short stylets with the vulva in the posterior of the body (**154**).

114 Dry rot of purple yam caused by *Pratylenchus coffeae* in Papua New Guinea.

Meloidogyne spp.

Distribution

A number of species of *Meloidogyne* have been found on yams, but the most commonly occurring is *M. incognita*. *Meloidogyne* spp. probably occur in all yam growing areas but have been reported on yams in west Africa, Pacific Islands, Caribbean, and Central, South, and North America.

Symptoms and diagnosis

The characteristic damage symptoms are galling of roots and also of tubers, which have the appearance of small to large warts over the surface (**115**). These tubers also tend to have an excess of roots growing from the galls. Necrotic spots associated with females also occur deeper in tuber tissues; *Meloidogyne* can penetrate deeper in tissues than do other nematodes, sometimes over 10 mm. Where this occurs, the females and egg masses normally become surrounded by necrotic cells as the host reacts. Root galling can reduce plant growth and cause foliar chlorosis and premature leaf fall.

Economic importance

Meloidogyne spp. can severely reduce growth and yield of yams and are particularly damaging in the early growth stages. The marketable value of tubers is greatly decreased because of the unattractive appearance of tubers and the presence of necrotic spots in the tissues.

Management

Most of the methods for other nematodes can be used for managing *Meloidogyne* spp., especially nematode-free seed tubers. Root knot nematodes have a wide host range making rotation a difficult proposition, but intercropping with highly susceptible crops such as tomato, okra, and beans should be avoided.

115 Galled tuber infested with *Meloidogyne incognita*.
(Courtesy of R.A. Plowright.)

Cassava *Manihot esculenta*

Cassava is a tropical crop originating in South America but now a very important food crop throughout the tropics, especially in Africa. Many nematodes are found with cassava but only *Meloidogyne* species and *Pratylenchus brachyurus* are known to be damaging pests, root knot being the most important. However, there are conflicting reports as to the extent of damage caused by *Meloidogyne* spp. on the crop. Cassava has been found to be both susceptible and resistant to root knot, due largely to the different biological isolates of the nematodes present in the growing region, but this can also be due to the host reaction of the many different types and cultivars of cassava being grown.

Meloidogyne spp.

Distribution

The main species of root knot nematodes on cassava are *Meloidogyne incognita* and *M. javanica*. They have been reported parasitic on the crop in Africa (Ivory Coast, Malawi, Mozambique, Niger, Nigeria, Uganda, South Africa), Brazil, Colombia, Venezuela, Fiji, Philippines, the USA, and India but are also associated with the crop in other parts of the world.

Symptoms and diagnosis

Cassava roots are naturally knobbly and slight nematode galling, when it occurs, is not always easy to observe unless healthy roots are present for comparison. Severely infested roots have obvious enlarged galls (**116**) sometimes containing as many as 80–100 swollen females/g root within the tissues, often surrounded by necrotic cells (**117**). The root systems are reduced and the swollen, storage roots normally found on healthy plants (**118**) are decreased or even absent on infected plants (**119**). Severe nematode infestation causes reductions in stem girth, height and weight, and is associated with death of plants in the field (**120**).

Economic importance

The nematode is only recognized as a problem in relatively few countries but, where it occurs, losses to storage root yield can be 17–50% or even as high as 98%. Stored roots infested with root knot can

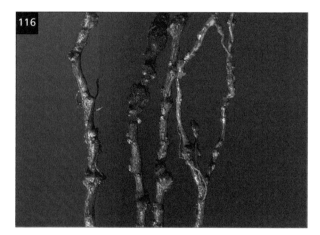

116 Root galls on cassava caused by *Meloidogyne incognita*.

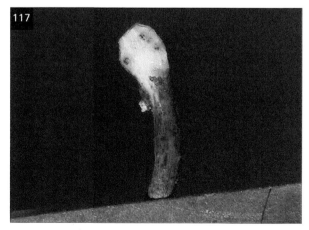

117 Section through cassava root gall showing *Meloidogyne incognita* females surrounded by necrotic cells.

118 Healthy cassava plant with swollen storage roots.

119 Cassava plant severely infested with *Meloidogyne incognita* without storage roots.

120 Dying cassava plant in a field heavily infested with *Meloidogyne incognita*.

deteriorate leading to considerable post-harvest losses. Cassava is propagated by stem cuttings and the other negative effects are related to the quantity and quality of the planting material available.

Management

Nematodes are not disseminated in stem cuttings and therefore they are primarily a field problem. Avoiding areas previously grown to cassava, and where root knot is known to occur in high populations, will reduce the damage. Use of resistant or immune cultivars, if available and suitable, is the most effective means of managing root knot and resistance has been identified in a number of countries.

Pratylenchus brachyurus

P. brachyurus is a migratory endoparasitic nematode retaining it vermiform shape throughout its life history. It is a very common species in the tropics and is reported parasitic on cassava in Nigeria, Uganda, Togo, Ivory Coast, Brazil, Fiji, Malaysia, and Florida. The nematode is reported to cause a gradual decline in yields as populations build up in the soils when the crop is grown on a continuous basis. Sequential cropping and use of resistant cultivars are probably the best means of managing the nematode.

Taro *Colocasia esculenta*

Colocasia is grown for its edible corm which is known throughout the world variously as taro, cocoyam, dasheen, or eddoe. It can be confused with the similar crop *Xanthosoma* which is also referred to as cocoyam in west Africa. The known major nematode pests of *Colocasia* are *Meloidogyne* spp., *Hirschmanniella miticausa*, and *Pratylenchus coffeae*. *P. coffeae* is known as a pest of taro in Japan although it has also been found on the crop in the Pacific Islands.

Meloidogyne spp.

Distribution
The root knot nematodes can be found on the crop wherever it is grown but have been reported associated with damage to taro in the Caribbean, Pacific, India, Taiwan, and west Africa.

Symptoms and diagnosis
Typical galling of roots can be observed but these are not always obvious; galls can occur on the corms. Infested plants have yellowed leaves and generally poor growth. Swollen females can be teased from the root and corm tissues, and infective juveniles and males can be extracted from the soil.

Economic importance
Meloidogyne spp. are associated with severe losses in India, and corm damage can reduce the market value of the crop.

Management
Management is mainly by use of nematode-free planting material and avoiding growing the crop in land known to be heavily infested with the root knot nematodes.

Hirschmanniella miticausa

Distribution
H. miticausa is the cause of a rot disease of taro corms, known locally in the Pacific Island countries as the pigeon English name of 'miti miti' named after the symptoms of the disease which resemble red meat (**121**). It has a limited distribution to date, having been found so far only in the Solomon Islands and in Papua New Guinea in the South Pacific but is likely to be more widely spread.

Symptoms and diagnosis
Rotting of the corms is the most important symptom that can be observed when the corms are cut open. In the more lightly infested tissues, red streaks can be seen towards the bottom of the corm (**121**); brown rot is seen in heavily infested corms (**122**) and these can be completely decayed. Chlorosis and wilting of the leaves of infested plants also occurs. The nematode is a large, migratory endoparasite mainly of taro corms (**123**) but can also be found in roots and soil.

Economic importance
H. miticausa can be the cause of serious, or even complete, loss of the crop especially in areas where taro is grown year on year.

Management
Use of nematode-free planting material in land in which *H. miticausa* is absent is the most economical and sustainable means of preventing damage. The planting material that is generally used is the top

portion of the corm and leaf bases. The more of the actual corm tissue that is removed to healthy, white tissues, the greater the chance of having nematode-free material.

Identification

Females are long and thin, almost 2 mm in length, males slightly shorter, both with pointed tails (**124**). The stylet knobs are short but prominent in both sexes (**125**) and the vulva in the female is in the mid body.

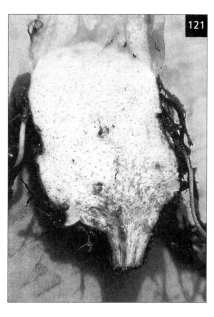

121 Taro corm with miti miti disease, showing red streaks and red necrotic areas caused by the feeding of *Hirschmanniella miticausa*.

122 Miti miti diseased taro corm with severe rot, infested with *Hirschmanniella miticausa*.

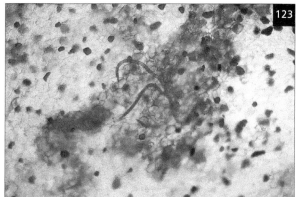

123 *Hirschmanniella miticausa* in taro corm tissues.

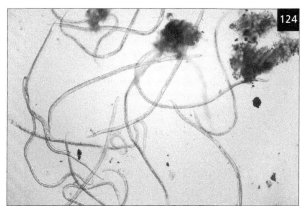

124 Females and male *Hirschmanniella miticausa* nematodes extracted from taro corm.

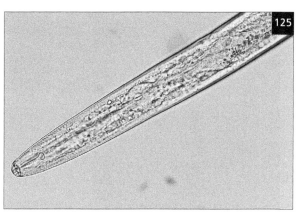

125 Anterior of *Hirschmanniella miticausa*.

Ginger *Zingiber officinale*

Ginger is grown as a spice crop for its underground rhizome throughout the world but particularly in Asia, Australia, the Pacific Islands, and parts of South America and Africa. The major nematodes damaging the crop are *Meloidogyne* spp., *Radopholus similis*, and *Pratylenchus coffeae*.

Meloidogyne spp.

The root knot nematode species that have been found associated with damage to the ginger crop are *Meloidogyne incognita*, *M. javanica*, *M. arenaria*, and *M. hapla*.

Distribution
The two most commonly occurring root knot species, *M. incognita* and *M. javanica*, occur as pests of the crop in most of the major ginger growing areas.

Symptoms and diagnosis
Meloidogyne produces characteristic galling of the ginger roots (**126**) but damage to the underground rhizomes is also very important. Rhizomes infested with the nematodes have lesions throughout the tissues (**127**) and unusual brown to black, fluid-filled lesions observable on the surface, each covered by an almost transparent layer of cells. These outer lesions are mainly found in the creases of the knobbly rhizomes and are very similar to those caused by the other nematode pest of ginger, *Radopholus similis* (**128**). Chlorosis and stunting or twisting of the foliage can also be seen. Seed rhizomes or pieces of rhizomes infested with the nematodes are the main source of infection in new land.

Economic importance
Yield losses by *Meloidogyne* species can be considerable, in excess of 50%. In addition, the marketable value of the rhizomes is greatly reduced which is of particular importance in the export trade.

Management
In large-scale, commercial production of ginger, such as in Queensland, Australia, chemical nematicides have been used but application has been declining. Alternatives to chemical control include the use of nematode-free planting material, avoiding growing successive ginger crops or other crops susceptible to root knot nematodes in the same land, rotating with a green manure crop (eg. forage sorghum, *Sorghum bicolor*), and application of large quantities of

126 Galling on roots of ginger caused by *Meloidogyne javanica*. (Courtesy of S.R. Gowen.)

127 Ginger rhizome infested with *Meloidogyne javanica* showing lesions around nematodes. (Courtesy of S.R. Gowen.)

organic amendments. Nematodes can be eliminated from seed rhizomes by dipping in hot water at 50–51°C for 10 minutes, but damage to the plants can occur if temperatures are allowed to vary.

Radopholus similis

Distribution
The burrowing nematode, *R. similis*, is a known pest of bananas and plantains but is also a serious pest of other crops including ginger. It is reported on ginger from Florida, India, Fiji, and Brazil.

Symptoms and diagnosis
The observable lesions caused by *R. similis* are similar to those produced by *Meloidogyne* spp. Liquid filled, shallow, brown to black lesions with an almost transparent cover occur on the surface of the rhizomes (**128**). Severe rotting can occur probably as a result of secondary pathogens,

128 Black watery lesions on the surface of rhizomes infested with *Radopholus similis* in Fiji.

129 Rotting rhizomes infested with *Radopholus similis* abandoned in a farmer's field, Fiji.

sometimes leading to loss of most of the crop in the field (**129**) accompanied by an unpleasant, pervading aroma. In the growing plant, nematodes cause stunting, chlorosis, and reduction in shoots. This migratory endoparasite can be found in the root and rhizome tissues and in the soil. The nematodes penetrate up to 1 cm deep in rhizome tissues, producing galleries.

Economic importance
Estimates put the yield loss at at least 40% where the nematode is prevalent and present in high populations in India and Fiji.

Management
As with *Meloidogyne*, use of seed rhizomes free from *Radopholus* is of paramount importance in managing the nematodes, combined with rotation of ginger with nonhost crops. In Fiji, *Xanthosoma* was found to be a nonhost to *Radopholus*.

Identification
There are no perceptible morphological differences between the *R. similis* that occurs on ginger and that occurring on banana and other crops (**143**). However, there are biological isolates or races of the species occurring with different hosts. The ginger isolate of *R. similis* from Fiji is not a pest of citrus, for example.

Pratylenchus coffeae

P. coffeae is very similar in biology to *Radopholus*, causing the same damage and lesions in the rhizomes. A small migratory endoparasite, the female is 0.5–0.7 mm long. It is found damaging ginger in India in the states of Kerala, Himachal Pradesh, and Sikkim. Hot water treatment can be effective in producing nematode-free seed rhizomes and crop rotation with nonhost crops should be employed. However, the geographical isolates of *P. coffeae* that have been studied have been shown to have a very wide host range and it is necessary for this information to be available for the ginger isolate from India before sequential cropping can be recommended. Other species of *Pratylenchus* are known from ginger, but the role of these as parasites of the crop is not understood.

Carrot *Daucus carota*

Carrot as a vegetable crop is parasitized by the same or similar wide range of nematodes discussed under vegetables (Chapter 3). These include, of paramount importance, the root knot species (*Meloidogyne hapla*, *M. incognita*, *M. javanica*, *M. arenaria*), and also the false root knot, *Nacobbus aberrans*, the sting nematode, *Belonolaimus longicaudatus*, lesion nematodes, *Pratylenchus* spp., and the stubby root nematodes, *Paratrichodorus* and *Trichodorus*. However, the symptoms, damage, and economic importance are related to carrot being a root crop.

Distribution
The tropical root knot nematode species are *M. arenaria*, *M. incognita*, and *M. javanica* and are found in hot tropical or warm climates; *M. hapla* is a nematode of the cool temperate countries and climatic regions such as at high altitude in the tropics. *Nacobbus* is also a temperate nematode with its origin in the Andes and is found in South and Central America and the southwestern USA. *B. longicaudatus* is restricted to the lower coastal plain of the southeastern USA. *Pratylenchus* species can be both tropical and temperate and are found world-wide. *Paratrichodorus* and *Trichodorus* are mainly temperate nematodes.

Symptoms and diagnosis
MELOIDOGYNE AND NACOBBUS
Root knot nematodes on carrot cause excessive root growth and galling (**130**), but even with extreme damage to the storage root the top growth can appear relatively normal (**131**). *M. hapla* produces small bead-like galls similar to those caused by *Nacobbus*, and often a mass of small roots giving the carrot a 'bearded' appearance.

BELONOLAIMUS LONGICAUDATUS
The sting nematode is the cause of necrosis and stunting of the developing storage roots. Aboveground, chlorosis and wilting are symptoms of severe infestation. The nematode is a relatively large ectoparasite with a long stylet (**132**) found in the soil and feeding externally to the roots; it has a 1 month life cycle.

PRATYLENCHUS SPP.
The migratory endoparasitic lesion nematodes cause necrosis in the developing storage roots. They can be extracted from the tissues and from the soil. Temperate species have longer life cycles than those from the tropics.

PARATRICHODORUS, TRICHODORUS
The stubby root nematodes stop normal growth of the storage root by feeding ectoparasitically on the root tip.

Economic importance
Nematodes, especially *Meloidogyne* species, are major pests of carrots and other vegetables. In addition to the yield losses caused, particularly at the young developing stage of the crop, the galled and distorted storage roots, often with excessive root growth, greatly decrease the marketable value of the produce.

Management
The methods available to manage the nematodes on carrot include rotating with nonhost crops, application of organic amendments, flooding, and biological control. Avoiding planting the crop in land known to be infested with any of the nematodes is the single most important means of reducing damage. All infested plant material should be removed from the land and not left to re-infest the soil and the following crop.

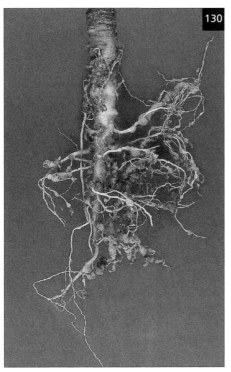

130 Galled carrot root and excessive root growth caused by root knot nematodes.

131 Extreme galling of carrot root caused by *Meloidogyne incognita*, with the top remaining relatively healthy.

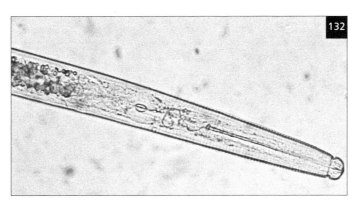

132 Anterior of female *Belonolaimus longicaudatus*.

Sugar beet *Beta vulgaris*

Sugar beet is an important crop in many parts of the world in both temperate and subtropical climates, and is therefore parasitized by a wide range of nematodes. It is extensively grown in western and eastern Europe and also grown in Asia, parts of South America, northern Africa, and Canada; in the USA, it is grown in seven western states. The most serious nematode pests of the crop are the beet cyst nematode, *Heterodera schachtii*, the root knot nematodes (*Meloidogyne* spp. both temperate and tropical), the false root knot nematode (*Nacobbus aberrans*), the stem and bulb nematode (*Ditylenchus dipsaci*), the stubby root nematodes (*Paratrichodorus* and *Trichodorus*), and the needle nematodes (*Longidorus* spp.).

Heterodera schachtii

Distribution
This cyst forming nematode is a very important pest of the crop, occurring in all the major sugar beet growing areas, especially in eastern and western Europe and the western USA. It is also present in limited areas of New York and Ontario where it is a problem on table beets, cabbage, and other vegetables. The species has also been found on table beet in west Africa.

Symptoms and diagnosis
Plants severely infested with *H. schachtii* are wilted under stress conditions (**133**), often seen in patches in the field (**134**). Invasion and feeding on roots by the infective juveniles causes new, secondary lateral feeder roots to grow producing a mass of extra small roots on the surface of the main, swollen root as it develops. The yield from the swollen root is reduced. The presence of nematodes can be confirmed by visual observation of white females on the surface of roots (**135**) or by extractions of active juveniles from the soil and roots, and extractions of lemon-shaped cysts (**136**) from soil; roots can also be stained and nematodes examined *in situ* microscopically.

Economic importance
The economic threshold for *H. schachtii* is 1–2 eggs/g soil, in sandy loam soils. Yield losses can be over 60% but are more likely to be 20–30% in mineral and organic soils. Continuous cropping on the same land is the cause of severe nematode population build up and yield loss.

Management
At least a 2-year rotation with root knot resistant tomatoes or fallow is recommended for effective control of *H. schachtii*. Metam sodium and aldicarb are the recommended nematicides with the nematode, but the use of pesticides is becoming less and less acceptable in many countries. Improved yields can also be obtained by planting at soil temperatures of <10°C for *H. schachtii* or <18°C for root knot nematode (see page 94), which allows early plant establishment prior to nematode activity. Plantings of green manure crops after sugarbeet, including mustard, radish, and rape, is also recommended as part of the management system. These act as catch crops and, if resistant varieties are used, they cause hatching of the nematode but do not allow full development in roots, thus reducing soil populations. Fungal biological control organisms also show promise for managing the beet cyst nematode.

Identification
As with all cyst forming nematodes, *H. schachtii* is identified mainly by the morphology of the cyst. It is distinguished from other cyst nematodes by features on the flask or lemon-shaped cyst (**136**).

133 Beet plants infested with *Heterodera schachtii* wilting in a field. (Courtesy of R.A. Sikora.)

134 Severe field damage to beet by *Heterodera schachtii* occurring in patches. (Courtesy of R.A. Sikora.)

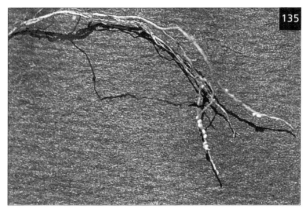

135 White females of *Heterodera schachtii* on beet roots. (Courtesy of R.A. Sikora.)

136 Brown cysts of *Heterodera schachtii* extracted from soils. (Courtesy of R.A. Sikora.)

Meloidogyne spp.

Distribution
The species of root knot found on sugarbeet and other beets in subtropical and warm temperate (Mediterranean) climates are mainly *M. incognita*, *M. javanica*, and *M. arenaria*. In temperate areas such as northern Europe and the northwestern USA, the other main species are *M. hapla*, *M. naasi*, and *M. chitwoodi*.

Symptoms and diagnosis
The galling characteristic of these species can be observed on the lateral roots. *M. hapla* and *M. naasi* produce relatively small galls but the tropical species, *M. incognita*, *M. javanica*, and *M. arenaria* cause large, pronounced galls. Root damage results in poor above-ground growth, chlorosis, and wilting, as seen with *H. schachtii* (**133**, **134**). The length of the life cycle of the different species depends on the temperature; the temperate species, *M. naasi* and *M. hapla*, normally only have one cycle on sugarbeet, whereas *M. incognita*, *M. javanica*, and *M. arenaria* in tropical soils have 4–5 cycles in each crop season. Root galling is the main means of distinguishing the problem, and types of galls can help in identifying the species causing the damage.

Economic importance
M. hapla, *M. naasi*, and *M. chitwoodi* are not as damaging pests in temperate countries. For the tropical species, yield losses of swollen roots can be over 50%.

Management
Rotation of crops is the most useful management means of reducing the damage caused by these species. However, as each *Meloidogyne* species has a different crop host range, it is vital that the species is correctly identified before a management programme is embarked upon.

Nacobbus spp.

Distribution
Two species of *Nacobbus* are known, *N. aberrans* (Thorne, 1935; Thorne & Allen, 1944) and *N. dorsalis* (Thorne & Allen, 1944). To date these are both only found as a pests of sugarbeet in North America. Most studies have been on *N. aberrans*.

Symptoms and diagnosis
Damage symptoms are, as the name implies, similar to those caused by the actual root knot nematodes, *Meloidogyne* spp. Severe root damage causes stunting and chlorosis reducing the yields of swollen roots. As in other crops, the galls are small and bead-like and spread along the roots (**100**).

Economic importance
Nacobbus spp. are of only local importance on sugarbeet in North America, but yield losses of over 20% have been recorded on the lighter soils in the USA.

Management
There are nonhosts of the nematodes which can be grown in rotation with sugarbeet to reduce populations and yield losses. There appears to be a separate sugarbeet race of the nematode in the USA and crops such as cereals, lucerne, onions, and even potato are reported to be nonhosts of this particular race.

Ditylenchus dipsaci

Ditylenchus dipsaci is a temperate nematode of limited distribution as a pest of sugarbeet in Europe and Switzerland and only a minor pest of the crop. *D. dipsaci* is a migratory parasite in which all stages are infective and foliar parts of the plant are parasitized. The fourth stage can survive in a desiccated state over winter or for longer periods. The symptoms of damage are variable depending on when and where the plant is attacked. Damage to the growing point of young plants results in the development of more than one crown, galls containing nematodes can form on the leaves, and petioles and leaves can be twisted. When the nematodes feed on the crown of older plants, a rot can occur.

To determine if damage is related to the presence of nematodes, *D. dipsaci* can be extracted or teased out from around diseased tissues. *Ditylenchus* spp. are not the easiest of nematodes to identify because they are very small and thin and have few readily observable distinguishing morphological characters. It is necessary to extract them from in or around diseased tissues to verify them as the causal organisms of any disease symptoms. The presence of a small stylet can separate *D. dipsaci* from the many bacterial feeding nematodes that occur in rotted tissues.

Management of the nematodes by nonchemical methods can be difficult because *D. dipsaci* has a very wide host range on crops and weeds.

Trichodorus spp., *Paratrichodorus* spp.

Species of these genera are pests of sugarbeet in Europe and the USA. In the UK, they are associated with poor growth of the crop known as 'docking disorder'. The main pest species are *Paratrichodorus teres*, *P. christiei* (= *P. minor*), *P. pachydermus*, and *Trichodorus primitivus*. These genera are migratory ectoparasites that feed mainly on the lateral roots, causing shortened or stubby roots. Damage to sugarbeet is localized and mainly restricted to very sandy soils. Symptoms of parasitism are stunted growth often in patches or rows related to short, stubby lateral roots below ground.

Trichodorus and *Paratrichodorus* differ morphologically from other parasitic nematodes in the shape of the stylet which is curved and lacking stylet knobs (**7**). They are short, sausage-shaped nematodes, 0.6–0.9 mm in length.

Longidorus spp.

The needle nematodes, mainly *L. attenuatus*, *L. elongatus*, and *L. macrosoma*, are only considered to be pests of sugar beet in the UK and eastern Europe. Feeding produces the formation of root tip galls that are the characteristic symptoms of damage to plants by these nematodes. Above-ground symptoms are stunted growth mainly confined to areas of sandy soils similar to that caused by *Paratrichodorus* and *Trichodorus* spp. Species of *Longidorus* have extremely long life cycles which can be as much as 1–2 years in the cooler, temperate soils. *Longidorus* spp. are the largest plant parasitic nematodes, with some species over 10 mm in length (**137**). They have very elongated stylets, often over 1 mm in length.

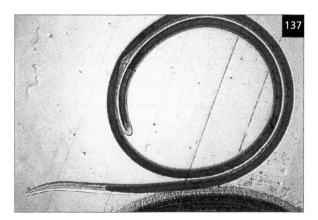

137 Female *Longidorus macrosoma*.

CHAPTER 7

Tree, Plantation, and Cash Crops

- INTRODUCTION
- BANANA AND PLANTAIN (*Musa* spp.)
  Radopholus similis; *Helicotylenchus multicinctus*; *Pratylenchus goodeyi*, *P. coffeae*; *Meloidogyne* spp.
- BLACK PEPPER (*Piper nigrum*)
  Radopholus similis; *Meloidogyne* spp.
- CITRUS CROPS
  Tylenchulus semipenetrans; *Radopholus* spp.; *Pratylenchus* spp.
- COCONUT (*Cocos nucifera*); OIL PALM (*Elaeis guineensis*)
  Bursaphelenchus cocophilus
- COTTON (*Gossypium* spp.)
  Meloidogyne incognita, M. acronea; *Rotylenchulus reniformis, R. parvus*; *Hoplolaimus columbus*; *Belonolaimus longicaudatus*

- TOBACCO (*Nicotiana tabacum*)
  Meloidogyne spp.
- COFFEE (*Coffeae* spp.)
  Meloidogyne spp.; Other nematodes of coffee
- SUGARCANE (*Saccharum officinarum*)
  Meloidogyne spp.; *Pratylenchus zeae*; Other nematodes of sugarcane
- PINEAPPLE (*Ananas comosus*)
  Meloidogyne javanica, M. incognita; *Rotylenchulus reniformis*; *Pratylenchus brachyurus*
- DECIDUOUS FRUIT AND NUT CROPS
  Meloidogyne mali, M. partityla; *Pratylenchus penetrans, P. vulnus*; *Criconemoides curvatum, C. xenoplax*; *Xiphinema* spp.

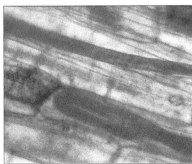
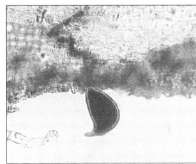
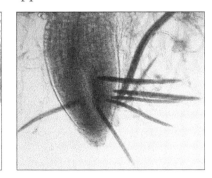

Introduction

The crops covered in this chapter are banana and plantain, black pepper, citrus, coconut, oil palm, cotton, tobacco, coffee, sugarcane, pineapple, and deciduous fruit and nut.

Banana and plantain *Musa* spp.

Nematodes are major root pests of banana and plantain and related *Musa* spp., with some nematode species causing very serious yield losses wherever they occur. The main nematode pests of these crops are the root parasitic species *Radopholus similis*, *Pratylenchus coffeae*, *P. goodeyi*, *Helicotylenchus multicinctus*, and *Meloidogyne* spp. They are mainly disseminated, both locally and globally, in infected planting material.

Radopholus similis

Distribution

The banana burrowing nematode, *R. similis*, is very widely distributed in most, but not all, tropical banana growing countries. It is of particular economic importance in commercial plantations in South and Central America, the Caribbean, parts of Africa, Australia, Pacific Islands, and some countries in south and southeast Asia.

Symptoms and diagnosis

R. similis is the cause of disease symptoms known as 'blackhead', 'root rot' or 'toppling disease'. Infested banana plantations have incomplete leaf cover and sparse growth (**138**). Individual plants are stunted with small bunches, few followers (suckers), and sometimes with yellow leaves; uprooting or toppling due to root breakage is common particularly after high winds (**139**). Exposed roots of toppled plants are blackened (**140**). When roots are examined internally, they have a characteristic purple to black necrosis throughout the root cortex but stele or

vascular tissues remain white in actively growing roots (**141**). Banana corms at root attachment points have similar necrotic patches where nematodes have invaded from roots (see **152**).

R. similis is a migratory root endoparasite found mainly around necrotic areas in root and corm tissues (**142**) but also in soil samples. The presence of nematodes in the crop is most easily verified by extraction from chopped or macerated roots showing some necrotic symptoms. A less time-consuming and more direct, practical approach to determining the presence of nematodes without having to resort to microscopic examination is by slicing the roots and recording the percentage of roots showing characteristic cortical necrosis on plants in the field.

Economic importance

Most yield loss estimates have been based on the increase in harvested bunch weights following the application of nematicides compared to nontreated plants. Where *R. similis* is the only important nematode present, these yield increases in banana and/or plantain can vary from 13–263% as in the

Ivory Coast. Similarly in other banana and plantain growing countries around the world, yield improvement after nematicide application has been recorded to be 20–40% in Cameroon, 35–40% in Madagascar, 86% in Panama, 5–30% in Australia, 38% in South Africa, over 70% in Ecuador and over 250% in St. Vincent and Puerto Rico.

Management

Commercial banana growers, especially those producing fruit for the international export trade, regularly use nematicides as pre- or post-plant treatments in the planting holes or around the established plants. A number of organophosphate and carbamate nematicides are used mainly as granular formulations. The cost of chemicals inhibits their use with most other banana and plantain growers, particularly small-scale and subsistence farmers. Alternatives are available and of particular importance

is the use of nematode-free planting material obtained from clean soils, or by paring corms or heat treatment (hot water or solarization) to remove diseased tissue or kill nematodes; tissue cultured plants used as planting material will also be free of nematodes.

138 Cavendish banana plantation (Ecuador) severely infested with *Radopholus similis*.

139 Toppled Cavendish bananas (Cameroon) infested with *Radopholus similis* after high winds.

140 Blackened and broken roots of toppled banana infested with *Radopholus similis*.

141 Necrotic lesions throughout the cortex of banana roots caused by *Radopholus similis*.

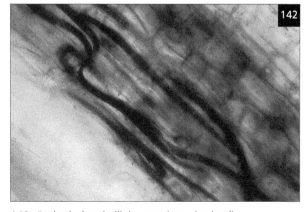

142 *Radopholus similis* in root tissue (stained).

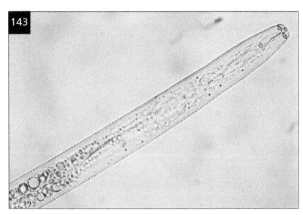

143 Female *Radopholus similis*.

tropical banana growing areas. On roots, small, rounded black lesions are first seen on the root surface leading to complete blackening of the outer parts of the roots. Internally, roots have shallow, black to brown necrosis in the epidermal and outer cortical tissues but not penetrating to the stele (**144**), in comparison to the symptoms caused by *R. similis* and *Pratylenchus* spp. *H. multicinctus* is a migratory endoparasite of the outer root cortex; all stages are also found in soil around roots. Root extraction is the easiest method of confirming presence of the nematode.

Economic importance
H. multicinctus is considered to be the main damaging nematode pest in areas which are sub-optimal for banana production where temperature and rainfall are limiting factors such as the Mediterranean and Middle East. Because of the lower temperatures, these are also areas where the generally more damaging nematodes such as *Radopholus* and *Pratylenchus* spp. are absent.

Other means of managing *R. similis* and other nematodes are: a break crop with a nonhost (eg. cassava) for 2–3 years, flooding land for 8 weeks after having destroyed the previous banana crop, fallow in absence of banana 'volunteers', and mulching with organic wastes. The propping or use of string guides (supporting with rope) of stems when bunches have formed reduces the risk of toppling and loss of bunches.

Identification
Females of *R. similis* are 0.5–0.8 mm in length, vermiform with tapering tails; short, robust stylet and rounded head; oesophageal region overlapping intestine dorsally; vulva in mid-body (**143**). Males are morphologically different from females.

Helicotylenchus multicinctus

Distribution
The banana spiral nematode, *Helicotylenchus multicinctus*, is probably the most commonly occurring nematode parasitic on banana roots and is found in both tropical and sub-tropical countries throughout the world. It often occurs together with other parasitic nematodes.

Symptoms and diagnosis
Reduced plant growth and bunch size are seen as a gradual decline in production over a number of years, sometimes accompanied by nutrient deficiency symptoms. Damage symptoms are more apparent on bananas grown in fringe areas outside the main

144 Necrotic lesions in the outer cortex of roots caused by *Helicotylenchus multicinctus*.

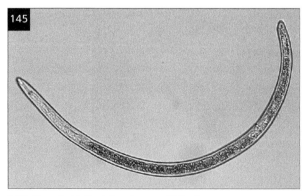

145 Female *Helicotylenchus multicinctus*.

Management

The methods described for *R. similis* apply although nonhost break crops have not been identified for *H. multicinctus* and the nematode causes relatively little toppling.

Identification

Females are 0.5–0.7 mm in length, lying in a 'C' shape when killed (**145**). The stylet is well developed, head is conical; oesophageal glands overlapping the intestine, mainly on ventral side; vulva is in the mid to posterior part of the body; the tail is short and rounded without projections.

Pratylenchus goodeyi

Distribution

Pratylenchus goodeyi, a lesion nematode, has a limited distribution compared to other banana nematodes, but is an important pest of the commercial banana crop in the Canary Islands, and of the banana food crop in Tanzania, Kenya, and Uganda in east Africa and Cameroon in west Africa; it is also reported from Crete. It is also a damaging pest of the closely related crop *Ensete* in Ethiopia. Its distribution is closely linked to cooler climates and thus at higher altitudes, the afromontane areas, in Africa.

Symptoms and diagnosis

Above-ground symptoms and root damage are very similar to those caused by *R. similis*. Growth and bunch size are severely reduced (**146**), toppling/uprooting is common; reddish to purple necrosis of roots extends throughout the cortex to stele (**147**, **148**) and necrosis of basal portion of corms occurs.

146 Cooking banana plantation (Tanzania) severely infested with *Pratylenchus goodeyi*.

147 Roots of toppled banana in Tanzania, broken and necrotic due to *Pratylenchus goodeyi*.

148 Necrotic lesions throughout the root cortex caused by *Pratylenchus goodeyi*.

P. goodeyi is a migratory endoparasite and all stages can be isolated from around necrotic tissues (**149**), also in soil around the roots. Determining the presence of nematodes and assessment of populations is as for *R. similis*.

Economic importance

P. goodeyi is only found as a pest of commercial dessert bananas in the Canary Islands and Crete, otherwise it is a major pest exclusively of small-scale cultivation of banana and plantain food crops in Africa. In the banana growing area of Tanzania, it has been found in over 90% of farms, associated with toppling and other damage symptoms in over 80% of farms.

Management

Chemical treatments have been used where *P. goodeyi* occurs on commercial dessert bananas in the Canary Islands, but are not appropriate for use on banana and plantain food crops in Africa. Other management methods listed above for *R. similis* can be used.

Identification

The female is slightly curved when heat killed, 0.6–0.7 mm in length, with a short tapering tail (**150**). It is similar in appearance to *R. similis* but with a posterior vulva.

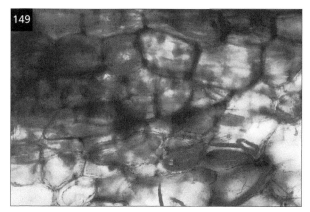

149 *Pratylenchus goodeyi* in root tissue (stained).

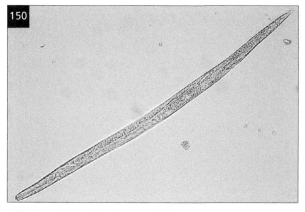

150 Female *Pratylenchus goodeyi*.

Pratylenchus coffeae

Distribution

P. coffeae, another lesion nematode, is recorded as a pest of bananas and plantains and other *Musa* spp. in Brazil, Ecuador, Brunei, the Canary Islands, Dominican Republic, India, Philippines, Seychelles, Papua New Guinea, South Africa, Thailand, West Indies, Costa Rica, Panama, and Honduras. It probably originates from the Pacific Island and Pacific Rim countries.

Symptoms and diagnosis

Damage symptoms are often undistinguishable from those caused by *R. similis* and *P. goodeyi*, although in some countries it is recorded that *P. coffeae* causes more severe damage to plantains, abaca (*Musa textilis*), and cooking and diploid bananas than to commercial dessert bananas. Toppling and uprooting occurs, root lesions are purple to black extending throughout the root cortex and extensively on corms (**151**) with necrotic patches at root attachment points (**152**).

P. coffeae is a migratory endoparasite, causing damage as it migrates through root and corm tissues. All stages, including eggs, are found around necrotic tissues (**153**) and also in soil around plants. Assessment of populations and damage is as for *R. similis* and *P. goodeyi*.

Economic importance

This species of *Pratylenchus* is especially important on bananas, plantains, and abaca (*Musa textilis*) in the Pacific Island countries, Central and South America, and Cuba where high populations occur. It causes toppling and is a limiting factor in production. It is also present as an economically important pest on dessert bananas in South Africa causing crop losses as high as 80% where it occurs.

Management

See *R. similis*.

151 Necrosis of root cortex and corms of banana caused by *Pratylenchus coffeae*.

152 Necrotic lesions in banana corm at root attachment points caused by *Pratylenchus coffeae*.

153 Stained *Pratylenchus coffeae* in root tissues.

Identification

Females are slightly curved when heat killed, 0.5–0.7 mm in length, with short tapering tail and posterior vulva (**154**). Distinguishing species of *Pratylenchus* is difficult and requires the help of an experienced taxonomist.

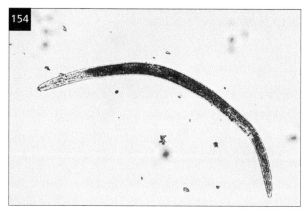

154 Female *Pratylenchus coffeae*.

Meloidogyne spp.

Distribution

The root knot nematodes, *Meloidogyne* spp., mainly *M. incognita*, *M. arenaria*, and *M. javanica*, are found world-wide on bananas and plantains.

Symptoms and diagnosis

Meloidogyne spp. do not produce any characteristic above-ground symptoms, apart from occasional reduced growth. There are few reports of them causing serious economic damage in mature plants but they can be a problem in young tissue culture seedlings. They cause root swellings or galls mainly on root tips of smaller roots (**155**), sometimes producing necrotic spots in root tissues when developing deep inside roots (**156**). *Meloidogyne* spp. are sedentary endoparasites, females becoming swollen when mature. Galling of roots is a clear indication of infection by *Meloidogyne* spp.

Economic importance

Although root knot nematodes are major pests on most other tropical and sub-tropical crops, they are generally ranked as minor pests on bananas and plantains except in a few specialized growing conditions such as under plastic in Morocco and Tunisia. There are also reports that *Meloidogyne* spp. can be damaging on meristem cultured banana plantlets used for propagation.

Management

Meloidogyne rarely occurs in the absence of other nematodes and management of this nematode alone is normally not justified. Ensuring that the soil in which banana plantlets are grown prior to planting in the field is free of nematodes will prevent damage during the early stages of growth.

Identification

Mature females are swollen and rounded (**157**) with a protruding neck, 0.5–1.0 mm in length; developing stages in roots are sausage-shaped, without stylets. Infective second-stage juveniles are vermiform, 0.35–0.5 mm in length, with a strong stylet and tapering, almost pointed tail (**158**).

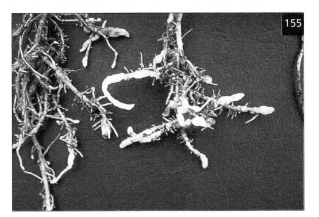

155 Galls on banana roots caused by the root knot nematode *Meloidogyne incognita*.

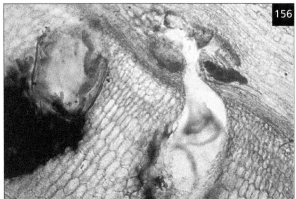

156 Female and eggs of *Meloidogyne javanica* deep in banana root tissue.

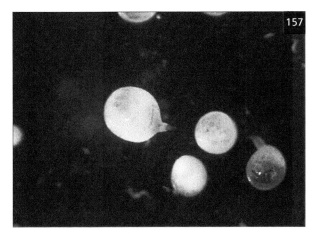

157 Swollen females of *Meloidogyne incognita*.

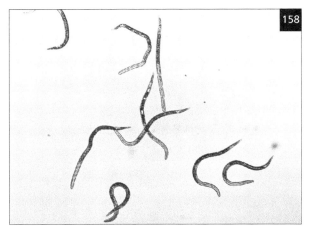

158 Second-stage infective juveniles of *Meloidogyne incognita*.

Black pepper *Piper nigrum*

This tropical perennial climbing shrub has its origins in Asia but is now cultivated in many other parts of the world where it encounters many nematode pests, although only two species, *Meloidogyne incognita* and *Radopholus similis*, are recognized as major pests of the crop.

Radopholus similis

Distribution
The isolate of *R. similis* recognized as the major causal organism in 'yellows disease' or 'slow wilt disease' of black pepper is found in India, Indonesia, Sri Lanka, Malaysia, and Thailand.

Symptoms and diagnosis
Progressive yellowing of the foliage (**159, 160**) eventually resulting in leaf fall and death of the vines (**161**) are the main symptoms of yellows (slow wilt) disease. In large plantations these yellowing and dead plants can be as spreading patches with apparent healthy vines on the fringes of the patches (**161, 162**). The migratory endoparasitic nematode invades the cortical tissues of black pepper roots causing initially necrotic, purple lesions followed by general rot. Nematodes can be extracted from both soil and roots, the latter being the most useful for confirming the causal organism of damage.

Economic importance
The nematode is infamous for being associated with almost the complete collapse of the black pepper industry in parts of Indonesia, destroying more than 20 million plants over a 20 year period on one island alone.

Management
There are no successful, nonchemical methods for managing nematodes on the standing perennial crop; chemicals are not considered an option by growers. The amount of yield loss and onset of the disease can be reduced by the use of mulches, addition of fertilizers and a good water supply. Use of healthy planting material when setting up a plantation is recommended.

Identification
R. similis females are vermiform, 0.5–0.8 mm in length (**143**).

Meloidogyne spp.

Distribution
The root knot nematodes (mainly *M. incognita* and *M. javanica*) have been reported on black pepper from Asia, southeast Asia, and South America.

Symptoms and diagnosis
The above-ground symptoms are poor growth and some discoloration of the leaves and wilting due to moisture stress. Root galling is obvious in heavily infected plants. Second-stage infective juveniles (**158**) and males are found in the soil; all other stages are found in roots.

Economic importance
Growth can be markedly reduced (up to 50%) in the presence of root knot nematodes; the reduction is directly related to nematode population levels in the soil.

Management
Use of healthy vines for planting is recommended. Some resistance to *M. incognita* has been detected in cultivars grown in India.

159 Early symptoms of yellows disease on black pepper vines in Bangka, Indonesia caused by *Radopholus similis* with apparent healthy vines in background.

160 Severe yellowing of black pepper caused by *Radopholus similis* in Bangka, Indonesia.

161 Dead and defoliated black pepper vines seen from the centre of infested patch caused by *Radopholus similis*, Bangka, Indonesia.

162 Spreading patch of yellowed black pepper vines in a plantation in Bangka, Indonesia.

Citrus crops

World-wide, the only widespread, economically important nematode on citrus is *Tylenchulus semipenetrans*, known as the citrus nematode. Others nematodes such as *Radopholus* and *Pratylenchus* spp. can be locally important.

Tylenchulus semipenetrans

Distribution
The citrus nematode is now present in most citrus growing regions of the world, having been disseminated on rootstocks.

Symptoms and diagnosis
The disease caused by *T. semipenetrans* is referred to as 'slow decline', which describes the symptoms on citrus trees. There is a gradual decline in growth, leaf size, fruit size, and yield over a number of years as the populations of the nematode increase in the soil and on the roots. Chlorosis of the leaves and twig die-back with leaf loss can also be observed, mainly on the upper branches (**163**). Damage symptoms can occur more rapidly when young trees are planted in land already heavily infested with the nematode. The nematodes produce sticky egg masses on the surface of the roots which become coated with soil particles, and these 'dirty roots' can be distinguished from the clean, healthy roots (**164**).

T. semipenetrans is a sedentary, semi-endoparasitic nematode. Second-stage juveniles are the infective stage. They are difficult to identify correctly, even at high magnification, but are all the same or similar size and usually occur in high numbers in soil extracts, which is a good clue to their identification. Juveniles partly invade roots and develop to mature females which protrude from the roots, and the exposed part of the mature

163 Twig dieback on citrus associated with severe infestation by the citrus nematode, *Tylenchulus semipenetrans*, in Malawi.

164 'Dirty roots' of citrus, a symptom of roots being infested by *Tylenchulus semipenetrans*.

female body becomes enlarged on the surface of the root (**165**). The presence of *T. semipenetrans* is best determined by examining the roots. The mature females are very small, less than 0.3 mm, and can normally only be observed when stained (**166**).

Economic importance

Yield decline is directly related to populations of *T. semipenetrans* in the soil, with an economic threshold level around 1000 juveniles/100 cm^3 soil. However, severely damaged roots of trees in terminal decline cannot support such large populations of the nematode. Generally, once decline of trees has begun, less time and input to manage these trees are used, and this neglect accelerates the decline in growth and yield.

Management

As citrus is a perennial tree crop, the use of nematicidal chemicals has been the main means of managing the nematodes in an established crop. The only other realistic means of preventing damage by the citrus nematode is by avoiding the problem by use of nematode-free rootstocks planted in nematode-free soil. Hot water treatment can be used to eradicate nematodes from the roots of rootstocks.

Identification

Juveniles are vermiform and are very small, 0.3 mm long, with small, delicate stylets only 10 μm in length. Swollen females are generally curved on the surface of roots (**165**) but remain very small, only 0.35–0.4 mm long. It is extremely difficult to remove them totally intact from the root.

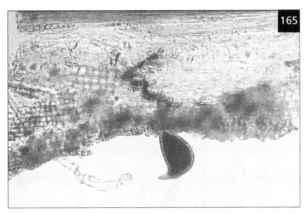

165 Section through citrus root showing female *Tylenchulus semipenetrans* as a semi-endoparasite embedded in the root cortex, stained with methyl blue.

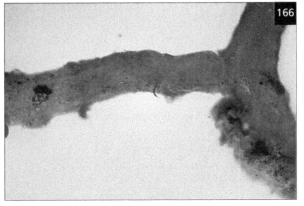

166 Citrus root with *Tylenchulus semipenetrans* females, juvenile, and eggsacs stained in acid fuchsin.

TYLENCHULUS SEMIPENETRANS

Radopholus spp.

Two species of *Radopholus* are known to damage citrus seriously, but both are very localized. *R. similis* (previously named *R. citrophilus*) is restricted to the centre of Florida in sandy soils. It is the cause of a disease known as 'spreading decline' whose symptoms are reduced fruit and leaf size and quantity. Twig die-back is normally associated with the nematode, and wilting can occur during dry spells. The *R. similis* populations on citrus are a biotype separate from those populations that occur on banana and other crops. The other species is *R. citri*, which has only been found in Sumatra, Indonesia. *R. citri* also causes stunted growth.

Radopholus spp. are migratory, endoparasitic, burrowing nematodes, causing cortical lesions and extensive root damage. Root systems invaded by the nematode are reduced and become dark brown resulting in poor growth (**167**). They are vermiform, 0.5–0.8 mm in length, with short but strong stylets and a vulva in the mid-body (**168**).

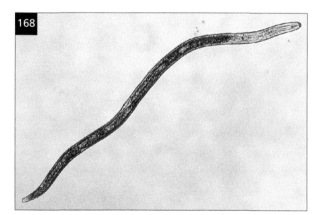

167 Citrus seedling infested with *Radopholus citri* (left) showing brown, necrotic roots and reduced growth compared to healthy seedling of the same age (right).

Pratylenchus spp.

As with other crops, many *Pratylenchus* spp. are found around the roots of citrus but the most damaging is *P. coffeae*. This species is found on citrus in many countries and continents including North America, Africa, India, and Japan. The species is a migratory, endoparasitic, lesion nematode causing similar root damage to that seen with *Radopholus* spp. The nematodes have a posterior vulva and short tapering tails, with a relatively prominent head and short, strong stylet (**154**). Diagnosis of both *Radopholus* and *Pratylenchus* spp. is best achieved by extraction of nematodes from roots showing symptoms of damage.

168 Entire female *Radopholus citri*.

Coconut *Cocos nucifera*
Oil palm *Elaeis guineensis*

The most well known and economically important nematode on coconut is *Bursaphelenchus cocophilus*, the cause of red ring disease. The only other nematode known to cause damage to coconut is *Radopholus similis*. Both these nematodes have a restricted distribution on coconut. Only the red ring nematode, *B. cocophilus*, is a pest of oil palm.

Bursaphelenchus cocophilus

The red ring nematode, now *Bursaphelenchus cocophilus*, was previously known as *Rhadinaphelenchus cocophilus*.

Distribution
B. cocophilus has only been found in the West Indies and in Latin America (Central and South America).

Symptoms and diagnosis
COCONUT
The internal symptom of nematode damage to the coconut palm is the characteristic ring of orange or red tissue seen in cross-section of the stem. The red ring is normally a few centimeters wide and some way in from the edge of the stem (**169**). In some coconuts, instead of a red ring there is more diffuse red necrosis of tissue throughout the stem (**170**).

169 Red ring disease of coconut caused by *Bursaphelenchus cocophilus* seen in cross-section of the stem, Belize.

170 Diffuse red necrotic tissue in coconut infested with *Bursaphelenchus cocophilus* seen in longitudinal section of stem, El Salvador.

The initial foliar symptoms of damage are yellowing of the leaves (**171**). Lower leaves of infested trees can be seen broken and hanging down and there is premature nut fall (**172**). Defoliation and eventual death is the fate of infested trees (**173**). The nematode is intimately associated with the palm weevil and sometimes the combined attack by the nematodes and insect can result in breaking of the stem (**174**).

OIL PALM

The ring of necrotic tissue caused by *B. cocophilus* in oil palm stems is less common than in coconut palm; where it occurs, it tends to be of a brown rather than red colour. The nematodes can be found in the crown of the oil palm infesting the young inflorescences, producing brown necrotic symptoms around the edges of the flower buds (**175**). The other symptom often associated with infestation by *B. cocophilus* is little leaf, which shows as short upright leaves in the growing tree (**176**).

B. cocophilus is an unusual plant parasitic nematode in that it has an insect vector, the palm weevil *Rhynchophorus palmarum*, that disseminates it from tree to tree. The palm weevil is also a pest of coconut, and its larvae feed on tissues within the stem (**177**). If the coconut is also infested with the red ring nematode they are picked up by the weevil

171 Initial yellowing symptoms on coconut infested with the red ring nematode, *Bursaphelenchus cocophilus*, Belize.

172 Coconut palms in El Salvador infested with *Bursaphelenchus cocophilus* showing leaves hanging down and premature nut fall.

173 Dead and dying coconut trees in Belize infested with the red ring nematode, *Bursaphelenchus cocophilus*.

174 Coconut tree damaged by both the red ring nematode, *Bursaphelenchus cocophilus* and the palm weevil, *Rhynchophorus palmarum* resulting in stem breakage.

larvae during this feeding process. The mature weevil that emerges transmits the nematodes to new trees via the ovipositor as it lays its eggs at the bases of leaf petioles in coconuts. The third-stage juveniles are the infective stage of the nematode.

Most biological studies have been done on the nematodes in coconuts. The life cycle in coconut is very short, often only 10 days, and very large populations of the nematode can be found in the diseased tissues. It may be necessary to actually cut down and destroy the palm tree to confirm presence of the red ring symptoms; an act that therefore must be performed with some conviction. Alternatively, a borer or sharpened steel tube can be used to take tissue samples from the stem; these tissues are macerated and nematodes extracted using Baermann funnels or their modifications.

Economic importance
Red ring is a major pest of coconuts and can cause severe losses if left unchecked. Once trees become infected they do not recover. In oil palm, the disease incidence is less.

Management
The destruction of diseased trees when symptoms become apparent is the most effective means of reducing the incidence and spread of the nematodes in both coconut and oil palm plantations. Trees can be cut down and burnt on the spot; or first treated with a herbicide to kill the trees before cutting them down and burning. Spraying with insecticides at bases of leaf petioles to kill the weevils is also recommended. Insect traps can be used in plantations to reduce the populations of the weevil vectors.

Identification
B. cocophilus are relatively long, vermiform nematodes with long tapering tails, around 1 mm for both females and males, but are also very thin. The vulva is positioned one-third of body length from tail tip. Stylets are small, 11–13 μm long and are often obscure. Morphological structures are not easy to observe.

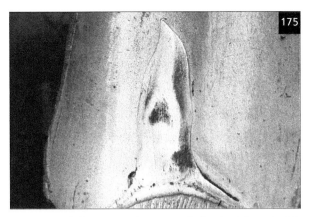

175 Young inflorescence in crown of oil palm infested with *Bursaphelenchus cocophilus* showing brown necrotic symptoms around the edges of the flower bud, Colombia.

176 Little leaf disease symptom in oil palm associated with *Bursaphelenchus cocophilus* showing shortened, upright leaves on the palm, Colombia.

177 Larvae and adult of the palm weevil, *Rhynchophorus palmarum*, taken from coconut stem.

BURSAPHELENCHUS COCOPHILUS

Cotton *Gossypium* spp.

Nematodes are economically important pests of cotton in all cotton production areas. Though there are four species of *Gossypium* grown commercially, *G. hirsutum* accounts for greater than 90% of total production. The most common nematode pests of cotton are *Meloidogyne incognita*, and *Rotylenchulus reniformis*. Nematode pests of lesser importance include *Belonolaimus longicaudatus*, and *Hoplolaimus columbus*. In some regions of Africa, *M. acronea* and *Rotylenchulus parvus* are also economically important.

Meloidogyne incognita and *M. acronea*

Distribution

Four host races of *Meloidogyne incognita* are recognized with races 3 and 4 being parasitic on cotton; these two races are commonly found in cotton production areas. In the USA, *M. incognita* occurs in every cotton growing area although severity of damage varies among these areas. Elsewhere, *M. incognita* has also been found on cotton in the Central African Republic, Ethiopia, Ghana, South Africa, Tanzania, Uganda, Zimbabwe, Brazil, El Salvador, Egypt, Syria, Turkey, Pakistan, China, and India. In areas where cotton has not been grown previously, native populations of *M. incognita* may not be parasitic on cotton. The other root knot species, *M. acronea*, is at present limited to Malawi and South Africa in southern Africa.

Symptoms and diagnosis

Affected plants are frequently stunted, chlorotic, and may senesce prematurely. Additionally, symptomatic plants are likely to occur in irregular clusters of varying size within a field, occurring especially in areas with more coarsely textured, sandy soils. At low to moderate nematode population densities or early in the season, root galling by *M. incognita* is usually indistinct with roots being only slightly swollen. With higher nematode population densities and near crop maturity, the galling of roots characteristic of root knot nematodes is more pronounced, but galls still do not reach the size seen on roots of other crops (**178**). Root galling is only slight with *M. acronea*, and enlarged females are frequently exposed and visible on the root surface; there is an increase in tertiary root growth around the nematodes (**179**). Fusarium wilt and seedling disease is typically more severe when plants are also parasitized by *Meloidogyne* spp. (**180**).

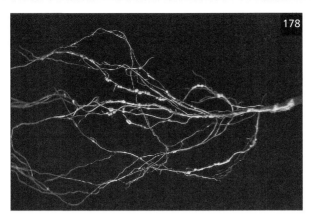

178 Typical galling on cotton roots caused by *Meloidogyne incognita*.

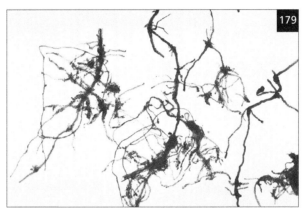

179 Slight galling of cotton roots and increased tertiary root growth around females of *Meloidogyne acronea* (stained) exposed on surface of root.

180 Increased Fusarium wilt of cotton in the presence of the root knot nematode *Meloidogyne incognita*.

Root galling is best evaluated by carefully digging up plants rather than pulling plants from the soil, which will strip off many of the lateral roots where galling is most prominent. The swollen, round females of *M. incognita* remain embedded in the root galls and brown egg masses are produced on the root surface (**181**). Because root galling by *M. acronea* is more limited, the swollen mature females usually rupture through the root cortex and are exposed on the root surface (**182**). Observation of females parasitizing roots or extraction of juveniles from soil and root samples are needed to confirm the diagnosis. Soil population densities of *M. incognita* are usually very low in the spring, and remain low until the latter half of the cotton growing season. Extraction of juveniles from egg masses by incubating infected roots on Baermann funnels can be used to confirm diagnosis early in the growing season when nematode populations in the soil are low.

Economic importance

Population densities of *M. incognita* as low as a single second-stage juvenile per 100 cm^3 soil at planting are sufficient to cause measurable yield suppression. Initial populations greater than 100 juveniles/100 cm^3 can result in complete crop failure. In the USA, losses are estimated to range from 1 to 5% of the total crop for individual states. In individual infested fields, losses of 10% are common. In some regions of the USA and Brazil, more than 50% of the fields are infested with *M. incognita* and, in these regions, losses in the absence of effective management can exceed 20% of the yield potential.

Management

In more developed cotton production regions, *Meloidogyne* spp. are managed with nematicides. Granular carbamates and organophosphates are most commonly used, followed by the fumigant 1,3-dichloropropene. Rotation with nonhost crops such as groundnut or with root knot resistant cowpea is also effective. Generally, 2 years of a nonhost crop between susceptible cotton crops is more effective than only a single year of the nonhost. A limited number of cotton cultivars with resistance to *M. incognita* is available. Many cotton cultivars are available with resistance to the Fusarium wilt/root knot complex, but these offer no resistance to the nematode itself. In the poorer farming communities of southern Africa, control of *M. acronea* could be achieved by rotating cotton with poor or nonhost crops such as millet (*Pennisetum* and *Eleusine* spp.), guar bean, or groundnut.

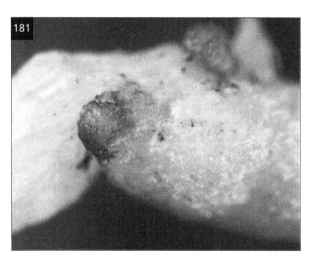

181 Brown egg masses of *Meloidogyne incognita* on surface of cotton root.

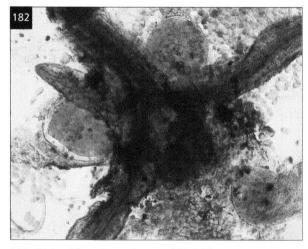

182 Stained females and eggs of *Meloidogyne acronea* on cotton root.

Rotylenchulus reniformis and R. parvus

Distribution

The reniform nematode, *Rotylenchulus reniformis*, is widely distributed in tropical and sub-tropical regions. The distribution of *R. parvus* is more limited, but it is found in several African countries, Australia, and the southern USA. These nematodes are rarely found in climates where soils are frozen for any lengthy period of time. The reniform nematodes can be found in soils of a wide range of textural classes but are commonly found in more finely textured, silty soils.

Symptoms and diagnosis

Reniform nematodes tend to be more uniformly distributed over a field than most nematode species and, hence, the field may lack irregular or discrete areas of symptomatic plants (**183**). The first symptom of damage may be a suppression of yield, followed by slight to severe stunting and chlorosis. Reniform nematodes can predispose cotton to seedling disease when initial nematode population densities are relatively high (>1,000 nematodes/500 cm^3 soil at planting). Roots of infected plants may be stunted, but lack diagnostic symptoms. Signs of the nematode include the presence of immature and mature females protruding from the surface of feeder roots (**184**) often obscured by egg masses (**185**) and, when egg masses are still present, they are generally covered with soil particles. When heavily infected roots are washed gently with water, the soil particles adhering to egg masses will give the roots a dirty appearance. Both soil and root samples should be examined for diagnosis. All life stages, except mature swollen females, can be extracted from infested soil.

Economic importance

Detectable yield suppression in a range of soil types can be observed when initial population densities of *R. reniformis* exceed 100 nematodes/100 cm^3 soil. Yield suppression of more than 50% has been observed when initial nematode populations exceed 1000 nematodes/100 cm^3 soil. Because of the widespread distribution of this nematode in the

183 Aerial infrared image of cotton field illustrating benefit of rotation to nonhost sorghum on left and relatively uniform distribution of damage caused by *Rotylenchulus reniformis* on the right. (Courtesy of C.M. Heald.)

184 Stained, swollen female of *Rotylenchulus reniformis* semi-endoparasitic in cotton root.

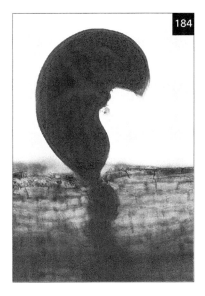

185 Mature female of *Rotylenchulus reniformis* on cotton root covered in gelatinous eggsac.

southern USA, annual total yield losses for cotton are estimated to be greater that US $20 million.

Management

Management has been primarily through use of granular and fumigant nematicide applications. Although *R. reniformis* has an extensive host range, rotation with nonhosts such as sorghum or groundnut provide effective management. Because of the ability of the nematode to survive desiccation, fallowing during a dry season is less effective than having the soil wetted periodically to induce nematode activity. No cotton cultivars with resistance to *R. reniformis* have been reported, but tolerant genotypes exist that have improved yields over susceptible cultivars. These tolerant genotypes, however, support high levels of nematode reproduction.

Hoplolaimus columbus

Distribution

The lance nematode, *Hoplolaimus columbus*, has a limited distribution, being found primarily in sandy coastal plain soils of the southeastern USA. It has also been reported from Egypt and Pakistan.

Symptoms and diagnosis

H. columbus feeds both as an ectoparasite and an endoparasite on the cortical tissues of cotton roots, causing large necrotic lesions. Affected plants are stunted with reduced root development, and typically occur in irregular clusters within the field (**186**). Suppression of tap root growth from nematode feeding may result in proliferation of lateral roots at the crown near the soil surface. Symptoms are similar to those caused by *Belonolaimus longicaudatus*. Diagnosis of the presence of nematodes can be made from either root or soil samples.

Economic importance

Yield suppression in the range of 10–38% has been documented in the southeastern USA when nematode population densities at planting were in the range of 90–200 individuals/100 cm³ soil. Total yield losses are much lower than for root knot or reniform nematodes because of the limited distribution of *H. columbus*.

Management

The primary management system is use of granular or fumigant nematicides. Although many common field crops are susceptible to *H. columbus*, rotations with non- or poor hosts such as peanut, sweet potato, tomato, peppers, and other vegetables are effective. Additionally, agronomic practices (soil fertility, irrigation, and subsoiling to disrupt hardpan layers in the soil) that reduce other stresses on cotton and promote greater root development are effective in reducing yield losses due to this nematode.

Identification

As with other members of the genus, *H. columbus* is a relatively large robust nematode, with mature females being 1.25–1.6 mm in length, with a heavily sclerotized and prominent head region (see *Hoplolaimus seinhorsti* **11**). They have a large, robust stylet (40–48 μm) with prominent knobs that resemble a tulip flower (**24**). The tail is blunt and the vulva is positioned near mid-body.

186 Damage to cotton in the field by *Hoplolaimus columbus*. (Courtesy of W.A. Powell.)

Belonolaimus longicaudatus

The sting nematode, *Belonolaimus longicaudatus*, is distributed throughout the southeastern USA, and has also been reported from California on turf grasses, and from the Bahamas and Bermuda, Puerto Rico, Costa Rica, and Mexico. Most surveys indicate that less than 2% of the cotton fields in the southeastern USA are infested with this nematode.

Symptoms and diagnosis

Symptoms caused by *B. longicaudatus* are similar to those of *H. columbus* and include primarily stunted, chlorotic shoots with poorly developed root systems (187). Sunken necrotic lesions in the root cortical tissues may be visible. Affected plants are typically distributed in irregular clusters throughout the field, with boundaries between healthy and stunted plants usually well marked. At high initial population densities (>100 nematodes/100 cm³ soil) seedling death may occur, resulting in reduced plant populations.

The sting nematode is a migratory ectoparasite, but has a long feeding stylet that allows it to pierce roots and feed on the cortical tissues (6). All life stages are parasitic. Distribution of this species is strongly influenced by soil texture and temperatures. The nematode is only found in the soil and requires very sandy soils (typically with greater than 85% sand content). During the warmest parts of the year, population densities may decline in the upper soil profile with most nematodes being found 15–30 cm deep.

Economic importance

Although *B. longicaudatus* is an aggressive nematode that can cause substantial plant damage where it occurs, it is of overall limited economic importance because of its restricted distribution. As few as 10 nematodes/500 cm³ soil at planting are sufficient to cause moderate damage to cotton.

Management

B. longicaudatus is most frequently controlled by nematicide application. Crop rotation with watermelon or tobacco is also effective. Peanut can be an effective rotation crop in those fields with populations of the nematode that are not parasitic on peanut.

Identification

B. longicaudatus is a relatively long (length 2–3 mm) and slender nematode, with a distinctly offset lip region and a long thin, flexible stylet (length 100–140 μm) with rounded knobs (6). The vulva is located near mid-body with two outstretched ovaries, and the female tail is comparatively long with a bluntly rounded terminus.

187 Field damage to cotton caused by *Belonolaimus longicaudatus*. (Courtesy of S.A. Lewis.)

Tobacco *Nicotiana tabacum*

Tobacco is grown extensively as a commercial nonfood crop throughout the world by both large and smallholder farmers, although demand for the product is on the decline. Without doubt, major pests of this crop are the root knot nematodes, *Meloidogyne* spp. Other nematodes are frequently found with the crop but are of minor importance or of only local importance. The tobacco cyst nematode, *Globodera tabacum,* for example, has been known to cause high yield losses in tobacco growing states of the USA.

Distribution

In the sub-tropics and tropics, the main economically important *Meloidogyne* species on tobacco are *M. incognita*, *M. javanica*, and *M. arenaria*. In the cooler temperate growing areas, *M. hapla* can also be important.

Symptoms and diagnosis

Root galling is the most obvious and characteristic symptom of infestation by the root knot nematodes if the plants are uprooted (**188**). Galling can be extreme where soil populations of the nematode are high. Foliar symptoms due to root damage are stunted growth, chlorosis, and wilting in dry conditions. These can occur in patches in the field or in individual plants scattered throughout the field (**189**, **190**) indicating infection of transplanted seedlings (**191**). The presence of the nematodes and their role in any poor growth symptoms can be verified by uprooting the plant and examining the roots for characteristic galling (**188**).

188 Galling of tobacco roots caused by the root knot nematode, *Meloidogyne incognita*, Malawi.

189 Poor and patchy growth and yellowing of tobacco in Bolivia severely infested with the root knot nematode, *Meloidogyne incognita*.

Economic importance

In the warmer tropical tobacco growing areas, root knot nematodes are major economic pests of the crop and it is rarely possible to grow the crop without managing the nematodes. Generally, if the tobacco species of root knot nematode is present in the soil, even at levels as low as 1 juvenile/100 cm^3 of soil, economic damage to the crop will occur. In the absence of nematode management, particularly with chemicals, losses can be over 50%.

Management

Chemical nematicides have been the favoured means of managing the root knot nematodes in field soil but the range of these nematicides available to tobacco farmers is now more limited.

Populations of root knot nematodes rapidly increase in soils under a very susceptible host crop such as tobacco. Rotation with nonhost or resistant crops is an effective means of reducing populations below an economic threshold level. Depending on the species and biological races of the nematodes present in the field soil, a range of nonhost crops can be grown. Maize and other cereals are generally effective against most of the root knots, groundnuts give good control of *M. javanica* and *M. incognita* but not of *M. arenaria*; cotton is a good rotation crop if the cotton race of *M. incognita* is not present. Pasture grasses are also effective. Nematode antagonistic plants such as sunnhemps, *Crotalaria* spp., also give good control. Nematode resistant cultivars are available to certain races of *M. incognita* and *M. javanica*.

Tobacco is grown as a transplanted crop and nematode infection can originate from the nursery beds. Ensuring that these beds are free of root knot nematodes is a vital part of nematode management in tobacco. Control of nematodes in nursery beds has, in the past, been achieved by the use of chemical nematicides including fumigants, such as methyl bromide, although many of these are no longer acceptable or permitted. Selecting sites that have not grown crops susceptible to root knot or have been flooded and are likely to be free of the nematodes is a practical alternative.

190 Individual stunted, chlorotic plants infested with the root knot nematode, *Meloidogyne incognita*, in tobacco field in Malawi.

191 Tobacco seedling prior to planting out in the field with root galls caused by *Meloidogyne incognita*.

Coffee *Coffeae* spp.

The two main coffee species grown as commercial plantation crops are Arabica coffee, *Coffeae arabica*, and Robusta coffee, *Coffeae canephora*, although other species are also grown on a small scale. Many nematode parasites are found around the perennial coffee plants but the most economically important world-wide are a range of different root knot nematodes, *Meloidogyne* spp. The only other recognized important nematode pests of coffee are species of the lesion nematodes, *Pratylenchus* spp., especially *Pratylenchus coffeae*.

Meloidogyne spp.

Distribution

Some of the *Meloidogyne* species or groups of root knot nematodes parasitizing coffee are restricted to only certain parts of the world at present, but could easily be disseminated to other coffee growing areas in future. *M. exigua*, which was the first nematode pest to be reported on coffee, is only found in Central and South American countries. Another South American species, *M. coffeicola*, only occurs in Brazil. A group of coffee nematodes known collectively as the African coffee root knot nematodes, *M. africana*, *M. decalineata*, *M. kikuyensis*, and *M. megadora* plus some undescribed species of the genus, only occurs in Africa (**192**). *M. konesis* on coffee is only known from the Hawaiian Islands. The common root knot nematodes, *M. incognita* and *M. javanica* are also found as major pests of coffee and these two species have a more widespread distribution on the crop and also a much wider host range.

Symptoms and diagnosis

The above-ground field symptoms are reduced growth (**193**), chlorosis, shedding of leaves, and general defoliation (**194**) and, in very serious infestations, death of the plants. Coffee seedling are particularly susceptible to damage and are also the means by which the nematodes are inadvertently disseminated (**195**, **196**). Root symptoms vary between species of *Meloidogyne*. The African coffee root knot nematodes and the South American species, *M. exigua*, produce small galls of characteristic shapes. *M. decalineata* causes small rounded galls to develop on the root tips often covering the whole root system (**197**). *M. exigua* also produces rounded galls but along the roots (**198**). *M. incognita* causes cracking and a breakdown of cortical root tissues resulting in dark, necrotic areas and death of portions of the root. Parasitism by *M. coffeicola* does not result in galls being formed but does produce cracking and breakdown of cortical tissues, mainly in the older roots in contrast to the other species which are mainly found in the younger roots.

192 Coffee infested with an African coffee root knot nematode, *Meloidogyne decalineata*, in the foothills of Kilimanjaro, Tanzania.

193 Stunted regrowth of coffee bush due to severe infestation by *Meloidogyne exigua* in Brazil.

194 Poor growth and defoliation of coffee plant severely infested with African coffee root knot nematode species, *Meloidogyne* spp., in Tanzania.

195 Coffee seedlings with root galls caused by *Meloidogyne decalineata* in Tanzania.

196 Coffee seedlings with root galls caused by *Meloidogyne exigua* in Peru.

197 Part of coffee root system from Tanzania covered in rounded root tip galls caused by an African root knot nematode, *Meloidogyne decalineata*.

198 Small galls on roots of coffee caused by the root knot nematode, *Meloidogyne exigua*, in Bolivia.

MELOIDOGYNE SPP.

Although most *Meloidogyne* species produce egg masses that are visible on the surface of root galls, the egg masses of *M. exigua* and also *M. decalineata* are laid within the cortical tissues and not on the surface of the roots as with most other species. Diagnosis of root knot nematodes is mainly by examining the roots for characteristic galling and other symptoms. Staining roots will determine the presence or absence of females within root tissues. Juveniles can be extracted from the soil around infected trees.

Economic importance

Coffee root knot nematodes are of major economic importance where they occur and can reduce yields to such a level that the crop is abandoned or replaced. Losses in South America alone are in the region of US$ hundreds of millions.

Management

In a perennial standing crop such as coffee, that has become infected with root knot nematodes, management is mainly limited to application of nematicides although their use is now seriously restricted. Biological control could be a possibility with fungal or bacterial nematode pathogens but this means has yet to be proven. Resistance is known in certain *Coffeae* species and these can be used as rootstocks with susceptible commercial varieties being grafted on to them. Preventing the introduction of the nematodes into field soil is likely to be the best strategy in most circumstances. As nematodes are mostly introduced into the field on infested seedlings (**195, 196**), eliminating nematodes from seedbeds and coffee nurseries by various means such as heat treatment, solarization, and selection of nematode-free soil can be the most economical method of managing the nematodes.

Other nematodes of coffee

Although many other nematodes are found associated with coffee, their economic importance is minor on the crop compared to the root knot nematode pests. The two other groups or species of nematodes that are known to damage coffee are the lesion nematodes, *Pratylenchus* spp., and the reniform nematode, *Rotylenchulus reniformis*.

The lesion nematode species *P. brachyurus* and *P. coffeae* (**154**) are found on coffee in South America, the West Indies, Africa, Asia, and Indonesia. They feed as migratory endoparasites and cause cortical necrosis in roots, the damage resulting in stunted growth and yellowing of the leaves. Management is again by nematicides in the standing crop and use of nematode-free seedlings to prevent introduction. *R. reniformis* is found on coffee in India, the Philippines, and the Pacific Island countries.

Sugarcane *Saccharum officinarum*

Sugarcane is a commonly grown commercial crop in tropical countries with many different plant parasitic nematodes associated with it world-wide. Of these, the species that are of the most importance belong to the genera *Meloidogyne* and *Pratylenchus*, although others are recognized and are a contributory factor in yield loss of sugarcane.

Meloidogyne spp.

Distribution

At least six species of root knot nematodes have been found parasitic on sugarcane. As with many other crops, the species *Meloidogyne incognita* and *M. javanica* are most common and are widespread in sugarcane growing regions, especially in sandy soils.

Symptoms and diagnosis

On sugarcane, *Meloidogyne* species feeding on roots produce only small galls or swellings (**199**) that are not always obvious. Reduced growth of stems can occur. The presence of the nematodes and their role in any poor growth symptoms can be verified by uprooting the plant and carefully examining the roots for characteristic galling, or by staining and dissecting the swollen females from the roots.

Economic importance

Root knot nematodes reduce the growth of the canes, the number of tillers and, as a consequence, the sugar yield.

Management

Sugarcane is a crop grown over a number of years on the same land by ratooning, and management of nematodes by rotating crops is normally not possible. Avoiding sandy soils for cultivating the crop is advisable. Some resistant or tolerant cultivars to root knot are available but only to single species. Chemical nematicides are used in some countries where the crop is grown on sandy soils.

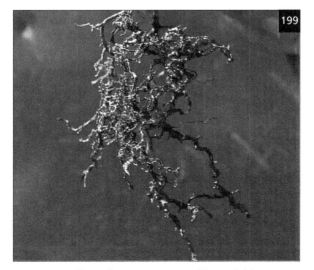

199 Root galling of sugarcane caused by *Meloidogyne incognita*.

Pratylenchus zeae

Distribution
Many species of lesion nematodes, *Pratylenchus* species, are found associated with the sugarcane crop but *P. zeae* is the most common and is found in a high percentage of sugarcane growing areas throughout the world.

Symptoms and diagnosis
The root symptoms are marked reddish lesions in the cortex and observed on the root surface (**200**). Root necrosis and rot causes a reduction in growth of canes. The endoparasitic lesion nematode can be extracted from both roots and soil, or observed within the roots after staining. *P. zeae* is a relatively small nematode, 0.4–0.6 mm in length, with a short but strong stylet 16 μm long (**8**, **82**).

Economic importance
P. zeae is second to the root knot nematodes in economic importance as a nematode pest of sugarcane.

Management
There are no practical alternatives to use of nematicides in the standing crop where populations of *Pratylenchus* spp. have become large. Resistance to this nematode is not available in sugarcane.

Other nematodes of sugarcane

Many other plant parasitic genera of nematodes are found in or around sugarcane roots. Ones that are considered of some importance as pests of the crop are *Hemicycliophora*, an ectoparasite (**201**), *Helicotylenchus*, *Hoplolaimus*, *Trichodorus*, *Paratrichodorus*, and *Xiphinema*. Others that appear damaging but have yet to be studied in detail can also be present in different parts of the world. One example is *Achlysiella williamsi* that has been found as a sedentary root parasite on sugarcane in Papua New Guinea and also reported from Australia. It is an unusual nematode with a swollen female body (**202**) that produces eggs in an egg mass on the surface of the root (**203**) and has a migratory immature female that is morphologically similar to *Radopholus* and is the infective stage (**204**).

200 Necrotic lesion on root of sugarcane due to feeding by *Pratylenchus zeae*.

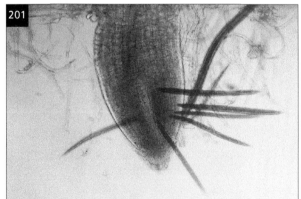

201 *Hemicycliophora* sp. feeding ectoparasitically on tip of sugarcane root.

202 Swollen adult female of *Achlysiella williamsi* from sugarcane root.

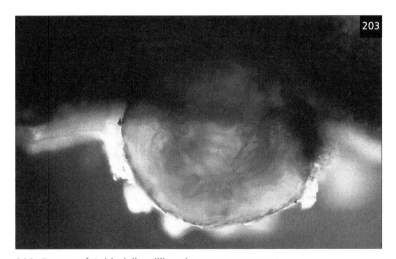

203 Eggsac of *Achlysiella williamsi* on sugarcane root.

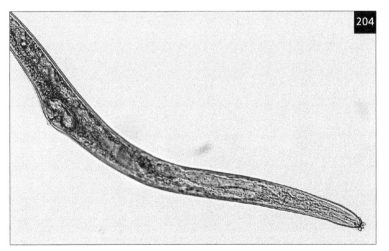

204 Immature female of *Achlysiella williamsi* beginning to swell.

Pineapple *Ananas comosus*

Pineapple is primarily a plantation crop with a total production of 14 million metric tons on 768,000 hectares in 2002. Asia (primarily the Philippines and Thailand) accounts for 60% of the world production, with 20% in the Caribbean, and Central and South America. The Ivory Coast, South Africa, Kenya, Australia, and Hawaii account for most of the remaining production. Because pineapple has a relatively long production cycle (22 months) that allows for multiple generations of parasitic nematodes and has a shallow, sparse root system, nematodes are often responsible for substantial yield suppression and economic loss. The most important nematode pests of pineapple include the root knot nematodes, the reniform nematode, and the lesion nematode.

Meloidogyne javanica and *M. incognita*

Distribution
M. javanica is the most commonly reported species of root knot nematode on pineapple, causing severe problems in Queensland, Australia and in Mexico, southern Africa, Thailand, and the Philippines. *M. incognita* has been reported from numerous pineapple production areas but is associated with economic losses only in Puerto Rico and Mexico.

Symptoms and diagnosis
Symptoms caused by all *Meloidogyne* species are similar, with plants stunted and chlorotic with high levels of infection early in the crop cycle. Root galling is moderate to severe, varying with the level of infection. Infection of pineapple typically results in terminal, fusiform root galls that result in a stunting of the root system (**205**).

Economic importance
The widespread distribution of root knot nematodes is a major reason for their economic importance. *M. javanica* is generally considered to be more aggressive than *M. incognita*. On most susceptible crops, initial population densities of 1–50 eggs and juveniles/500 cm^3 soil are sufficient to suppress yield.

Rotylenchulus reniformis

Distribution
Rotylenchulus reniformis is most frequently found attacking pineapple in Asia, Hawaii, and the Caribbean. The related *R. parvus* is often observed on pineapple in southern Africa but is rarely associated with economic loss.

Symptoms and diagnosis
Reniform nematodes tend to be more uniformly distributed over a field than most nematode species and, hence, the field may lack discrete, irregular areas of symptomatic plants. The first symptom of damage may be a suppression of yield, followed by slight to severe stunting and distinct reddening of the foliage. As with other hosts, distinct root symptoms on pineapple are lacking but, using a dissecting microscope, females may be observed protruding from the root surface; mature females covered with an egg mass (**19, 184, 185**), which is usually also covered with soil particles, can be observed with a hand lens or dissecting microscope. All life stages except the swollen female are easily extracted from soil. *R. parvus* can be readily distinguished from *R. reniformis* (0.4–0.5 mm in length) based on smaller size of the immature females (0.2–0.3 mm for *R. parvus*).

Economic importance

Damaging threshold population densities for *R. reniformis* are typically much greater than those of root knot species. For many susceptible crops, initial population densities of 500–1000 nematodes/500 cm³ soil are needed to suppress yields.

Pratylenchus brachyurus

Distribution

Pratylenchus brachyurus is most common on pineapple in the equatorial production regions and is less common in production areas located at greater latitudes.

Symptoms and diagnosis

Lesion nematodes cause distinct necrotic root lesions (see **200**), which are typically dark brown to black in colour. With severe infection levels there can be substantial destruction of cortical tissues and a loss of secondary roots. If infection is severe enough to stunt shoot growth, then the foliage may have a chlorotic appearance. Soil samples alone may fail to indicate the presence of damaging population densities of these nematodes, and nematodes need to be extracted from roots.

Economic importance

Greatest yield losses of pineapple are observed in Hawaii, the Ivory Coast, Uganda, and Brazil. Population densities of several hundred nematodes per gram of fresh root weight are needed to suppress yields.

Management

As a relatively high cash value plantation crop, nematode management in pineapple production has relied primarily on nematicide application that can be applied via drip irrigation (**206**). It is critical to suppress initial nematode population densities sufficiently to allow establishment of a healthy root system. The fumigant nematicide 1,3-dichloropropene usually provides better suppression of nematode populations than do any of the nonfumigant nematicides. Clean fallows of several months duration between pineapple crops are more effective for management of root knot and lesion nematodes than for the reniform nematode, which has a greater long-

205 Galled and stunted root system of pineapple infected with *Meloidogyne incognita*.

206 Drip feed irrigation system and plastic mulch in Hawaii.

term survival capacity due to its ability to enter an anhydrobiotic condition. Recent research indicates promise for improved management through the use of selected cover crops during the intercrop period. Both sunn hemp (*Crotolaria junica*) and rape (*Brassica napus*) suppress populations of the reniform nematode, in part through stimulation of biological control activities. Sunn hemp and rape, however, are good hosts for root knot nematodes and cannot be used in fields infested by *M. javanica*. No host resistance to nematodes is available in pineapple.

Deciduous fruit and nut crops

Of the many deciduous fruit and nut crops, apple (*Malus domestica*, *M. sylvestris*), cherry (*Prunus avium*, *P. cerasus*), peach (*P. persica*), walnut (*Juglands california*, *J. nigra*, *J. regia*), and pecan (*Carya illinoensis*) are among the most widely grown. All of these crops have several important nematode parasites that are able to affect productivity and longevity of the trees. Additionally, these nematode parasites are often key components in disease complexes, resulting in even greater economic loss. Several species of the root knot (*Meloidogyne* spp.) and lesion nematodes (*Pratylenchus* spp.) are parasitic on these crops. Additionally, the ectoparasitic ring nematode (especially *Criconemoides xenoplax*) and the dagger nematode (*Xiphinema* spp.) are parasites of widespread economic importance. The perennial nature of these crops contributes to the difficulty in managing nematode parasites. The specific cases listed below are a few examples of the types of disease incited by nematodes on a wide variety of deciduous crops.

Meloidogyne mali on apple, *M. partityla* on pecan

Distribution

The root knot species, *Meloidogyne mali*, parasitic on apple, is known only from Japan, and is especially important in Honshu and Hokkaido. *M. hapla* and *M. incognita*, which have world-wide distributions, have also been reported from apple. *M. partityla* was originally described from pecan in South Africa, but it appears to have originated in the southern USA and was introduced into South Africa on infected pecan seedlings. *M. incognita* and *M. javanica* are also known to parasitize pecan. All four of the common *Meloidogyne* species have been reported from peach.

Symptoms and diagnosis

Root symptoms vary with the species of *Meloidogyne* and host. Root galling can be especially prominent on peach infected by *M. incognita* but indistinct on pecan infected by *M. partityla* (**207**). *M. partityla* primarily infects the most recently produced feeder roots and is seldom found on older, larger roots. Where root galling is indistinct, as with *M. partityla*, mature females are often exposed on the root surface (**207**). Shoot growth may be suppressed with severe infections, resulting in thin crowns, shoot tip die-back, and chlorosis. Fruits are smaller in size and total fruit number may be reduced. Premature tree decline and senescence may be observed.

M. mali and *M. partityla* differ from most of the more common *Meloidogyne* spp. by having only limited host ranges. The host range of *M. mali* includes maple (*Acer* spp.), Japanese chestnut (*Castanea crenata*), cherry, mulberry, grape, and other *Malus* spp. The host range of *M. partityla* appears to be limited to the closely related species of *Carya* along with walnut and hickory (*Juglans* spp.). The optimal temperature range for *M. mali* is 20–30°C, and it apparently completes only one generation per year in northern Japan. No data are available on the specifics of the life cycle of *M. partityla* other than the observation that in infested pecan orchards the population has a peak in April and another in September. These dates correspond to periods of root growth. The limited number of generations per year completed by these nematodes makes it more difficult to diagnosis problems based on populations of juvenile nematodes present in soil samples; indeed for *M. partityla*, juveniles are rarely found in soil samples. Diagnosis is best achieved by observation of mature females on feeder roots collected from a number of trees showing symptoms of damage. The adult females can be difficult to dissect from woody root tissues, but can be identified by a globose shape with diameter up to 1 mm (see **157**).

Economic importance

Few data on actual losses due to *Meloidogyne* spp. are available for most of the deciduous crops. Several states in the USA estimate losses on apple in the range of 1–5% annually. In orchards in Japan infested with *M. mali*, apple growth suppression of up to 40% has been documented. Losses for peach due to *Meloidogyne* spp. in the southern USA are typically estimated to be greater than losses for apple. One early study estimated that initial population densities of 5 *M. javanica* juveniles/g soil would result in severe stunting. Nematode

control in a severely infested peach orchard may result in a cumulative yield increase of 2500 kg/ha over 3 years. Root knot nematodes on peach in Libya are recognized as contributing to substantial tree death. Loss in pecan due to *M. partityla* are estimated to be in the range of 1–3% annually in the southern USA.

Management

For all of the various deciduous crops, use of planting material free of nematode infection is the most important consideration in avoiding substantial losses. Pre-plant applications of a fumigant nematicide is recommended when replanting into an established orchard site known or suspected to be infested with root knot nematodes. There are several peach rootstocks that are resistant to *M. incognita* (**208**) and the use of these rootstocks are an effective management tactic for that crop. Because orchards are a long term investment of resources and because there are limited options for nematode management after planting, it is critical that a proper assessment of the potential for nematode damage is made and an appropriate management strategy prior to planting is developed.

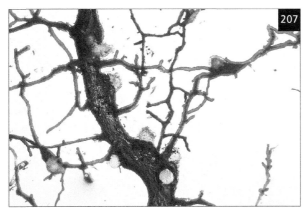

207 Roots of pecan infected with *Meloidogyne partityla*. Note the exposed females and limited galling.

208 Roots of *Meloidogyne incognita*-resistant Nemaguard peach and two root systems of susceptible Lovel peach from an orchard infested with the root knot nematode.

Pratylenchus penetrans and *P. vulnus*

Distribution

Of the numerous lesion nematodes, *Pratylenchus* spp., associated with deciduous crops, *P. penetrans* and *P. vulnus* are the most frequently encountered and cause the greatest amounts of damage. Both species have world-wide distributions, with *P. penetrans* being more common in regions with a temperate climate whereas *P. vulnus* is more common in the warmer climates of the Mediterranean region, Australia, and South Africa. The lesion nematodes are usually associated with well drained, sandy soils.

Symptoms and diagnosis

Infection of roots by all *Pratylenchus* spp. results in the formation of distinct necrotic lesions. The lesions are typically confined to the unsuberized feeder roots of deciduous crops. Root growth, especially of the feeder roots, is suppressed and the roots have a general discoloured appearance when nematode population densities are very high. There may be substantial reduction in occurrence of feeder roots. In severe cases, the plants suffer a general decline, with thin, chlorotic crowns. Tip die-back may also be observed. Nematode parasitism contributes substantially to complex disease syndromes, especially the peach tree short life syndrome and cherry decline.

The nematodes tend to accumulate in small colonies within the root tissues, with population densities of hundreds of nematodes per gram fresh root weight. Accurate diagnosis requires collection of root samples, especially of the feeder roots that are best collected from near the edge of the tree canopy. Nematodes can be extracted from these root samples by incubating the samples for at least 48 hours on Baermann funnels or in a mist chamber.

Economic importance

Few data are available of specific damage thresholds for any deciduous crop. Growth suppression of apple can be caused by 15 *P. penetrans*/100 g soil, 30/100 g for pear, 80/100 g for cherry, and 320/100 g for plum. In the Netherlands, 65% of the apple orchards were infested with *Pratylenchus* spp. One study found a 16% yield suppression of peach due to *P. vulnus*. The

greatest losses, however, are likely to be due to increased rates of tree mortality due to disease complexes such as cherry decline and peach tree short life syndrome, where stress caused by nematode parasitism increases susceptibility of the tree to other pathogens, such as bacterial canker caused by *Pseudomonas syringae*, and to winter freeze damage.

Management

Disease incidence and overall losses are usually greatest when replanting into old orchard sites where young trees may be exposed to high nematode population densities surviving from recently removed larger trees. Thus it is critical that the potential hazard is assessed and an appropriate management plan prior to planting is instituted. Little can be done to manage *Pratylenchus* spp. after planting. A single pre-plant treatment of each planting site with a fumigant nematicide typically improves rates of tree establishment, growth, and yield for several years. A fallow period of several months to 1 year can suppress nematode population densities; however, a clean fallow to avoid survival of the nematodes on alternate weed hosts must be maintained.

Criconemoides curvatum and *C. xenoplax*

Distribution

Criconemoides curvatum and *C. xenoplax*, commonly known as ring nematodes, appear to have nearly world-wide distributions, being reported from many regions of Africa, Asia, Europe, North America, and South America. *C. curvatum* and *C. xenoplax* cause the greatest losses to peach, almond, and walnut. The ring nematodes have been difficult to reconcile taxonomically as within the past 30 years the genus has variously been recognized as *Mesocriconema*, *Macroposthonia*, and *Criconemella*. Thus any search for additional information must include references to these multiple synonyms.

Symptoms and diagnosis

The typical symptoms of parasitism of roots of deciduous crops by the ring nematodes are a general decline in apparent root health, based on root biomass and colour. The abundance of feeder roots is diminished and the roots are discoloured, often due to the presence of necrotic tissues. Longitudinal cracks of older almond roots and necrosis of phloem tissues have been associated with parasitism by *C. xenoplax*. The shoots show various symptoms of decline and poor vigour, including some chlorosis, tip die-back, and thinner crowns. Nematode parasitism is frequently associated with complex disease syndromes, especially the peach tree short life syndrome. Bacterial canker caused by *Pseudomonas syringae* and increased winter freeze damage are components of the peach tree short life syndrome, and ring nematodes increase the susceptibility of peach and walnut to these maladies.

Criconemoides are ectoparasites with an almost sedentary life style. Although all life stages are motile, the nematodes may feed at a single location for several days. All stages feed and reproduction is probably by parthenogenetic mechanisms as males are rarely observed. Females lay eggs singly in the soil. The life cycle can be completed in about 4 weeks at 25°C. The nematodes can be readily extracted from the root zone soil by a variety of techniques; however, because of their sluggish movement, modifications of the Baermann funnel technique typically result in lower estimates of the population density than does extraction based on elutriation and/or sugar flotation.

Economic importance

The damage threshold for *C. xenoplax* on peach in sandy soils in the southeastern USA is 50–100 nematodes/100 cm^3 soil. Pre-plant fumigation with a broad spectrum soil fumigant in a site infested with *C. xenoplax* and prone to peach tree short life syndrome increased total yields over a 3 year period by more than 10,000 kg/ha. More than 40,000 ha of almond in California are estimated to suffer losses due to the direct and indirect effects of parasitism by *C. xenoplax*.

Management

As with other nematodes parasitizing these crops, diagnosis of the potential hazard prior to plant is a key to the reduction in losses. Post-plant application of nonfumigant nematicides such as aldicarb or phenamiphos provides better suppression of populations of *C. xenoplax* than it does for *Meloidogyne* or *Pratylenchus* spp. Pre-plant

fumigation is generally the optimal treatment. Reduction of other stress factors, such as those caused by mechanical weed cultivation, along with proper irrigation scheduling and fertility treatments, will reduce accompanying losses from disease complexes.

Identification

The *Criconemoides* are thick bodied nematodes of moderate length (0.40–0.60 mm) with a long stylet (71–86 µm), with sluggish movement, and are characterized by prominent annulations (2). A single ovary is present with the vulva located at 90% of the body length.

Xiphinema spp.

Distribution

There are seven *Xiphinema* spp., known as dagger nematodes, that are commonly associated with deciduous tree crops, including *X. basiri*, *X. diversicaudatum*, *X. vuittenezi*, and the *X. americanum* group (which also includes *X. brevicolle*, *X. californicum*, and *X. rivesi*). The combined distribution of these species on deciduous crops is world-wide. They are most commonly associated with apple, cherry, peach, and plum.

Symptoms and diagnosis

Roots of parasitized plants typically have swollen tips, often with necrotic lesions, and an overall reduction in root mass, although there may be a proliferation of lateral roots immediately behind a root tip that has been severely damaged. Shoots show symptoms of general decline, early leaf and fruit drop, and fruits may be reduced in size, and lack typical flavor. Parasitism by *Xiphinema* spp. is often accompanied by symptoms of the NEPO viruses these nematodes transmit (tomato ringspot virus, cherry rasp leaf virus, peach rosette mosaic virus, and strawberry latent ringspot virus). In addition to foliar symptoms of viral disease, there may be stem pitting of peach and, if the root stock is susceptible and the scion is resistant to the virus, there may be necrosis of the graft union such that girdling of the tree occurs, followed by tree death.

Xiphinema spp. (females 1–3 mm in length) are migratory ectoparasites that typically feed on the root tips, with all four juvenile stages and the adults being parasitic similar to *Longidorus* (see 5). They are characterized by long life cycles, typically requiring 1 year from egg to egg for *X. americanum* and up to 3 years for *X. diversicaudatum*. These nematodes may achieve soil population densities in the range of hundreds of individuals per 100 cm^3 soil and are readily extracted from soil. Highest populations can be expected from the areas of maximum root density, which is typically under the edge of the tree canopy.

Economic importance

Because of the importance of these *Xiphinema* spp. as virus vectors, it has been difficult to assess the independent impact on the nematodes on orchard health. A population density of 100 *X. americanum*/100 cm^3 soil can suppress the growth of peach seedlings. Several surveys in various regions of Europe and North America have indicated that nearly 90% of the orchards sampled were infested with one or more *Xiphinema* spp. The impact on orchard health was greatest where a NEPO virus was also present. Because of the additional damage caused by the viruses, an economic threshold of 1 nematode/100 cm^3 soil is often recommended for situations where both the nematode and a virus are present.

Management

Xiphinema spp. can be managed by pre- and post-plant applications of nematicides, with fumigant nematicides being the most effective pre-plant treatment. Incorporation of rape residues as a green manure is effective in suppressing population densities. Because of the nematode's sensitivity to dry soils, frequent cultivation prior to planting will suppress nematode populations in the shallow soil layers. It is also necessary to control the incidence of NEPO viruses. The first step is to use only certified, virus-free plant material when planting new trees. Some rootstocks and/or scions have resistance to important viruses and one should consult local authorities for availability of varieties with necessary resistance. The NEPO viruses have alternate broad-leaf weed hosts, such as dandelion, and are spread over long distances by the seeds of these weeds. Thus management of broad-leaf weeds in the orchard by cultivation or herbicides is important to prevent introduction of the viruses. Nematode transmission is the only mechanism by which the virus can move from the weeds to the orchard trees.

CHAPTER 8

Collection, Extraction, and Preservation of Plant Nematodes for Diagnosis

Introduction

The diagnosis of crop disorders caused by plant parasitic nematodes typically requires the collection of plant and soil samples to confirm the presence of nematode parasites. An accurate diagnosis is dependent on proper collection and processing of such samples. Improper collection and handling of samples may lead to the dismissal of nematodes as part of the problem, hence any management strategy developed to alleviate the problem will be deficient.

No diagnostician likes to receive an unidentified sample with the query 'Tell me what nematodes are present' or 'Did nematodes kill my crop?' Rather, it is important to send additional information to assist in the diagnosis of the problem. Such information should include, but not be limited to, the crop and cultivar, previous cropping history, history of other known or suspected problems, whether or not the field is irrigated or rainfed, and previous applications of soil amendments (organic or any pesticides). The source of the samples (town, district, county, state) is important. This information will permit comparisons with other problems reported previously from the regions, or indicate if the samples represent the first report of a nematode species from the area. It is impossible to provide too much information to the person making the diagnosis.

Collecting samples

Sampling field crops prior to planting to estimate a potential hazard requires a slightly different approach than sampling a current crop to diagnose an existing disorder. Plant parasitic nematodes are typically distributed in a nonuniform, nonrandom pattern. Rather, they tend to occur in aggregated clusters that are irregularly distributed across the field. Thus, estimates of nematode population densities are inherently imprecise and subject to a high degree of variability. To minimize such variability it is essential that soil samples be composites of multiple smaller samples. Numerous studies have shown that a sample composited from 20 sub-samples is an optimal in terms of minimising variation and optimising time and effort. There is little further reduction in variability of estimates of nematode densities with a

larger number of sub-samples, but there is a marked increase in such variability if fewer than 15 sub-samples are collected.

A second important consideration is the area to be covered by one sample. Variation in the estimates of nematode densities will increase with increasing sample area. A good compromise between variability and effort is to collect one sample from an area of no more than 5 hectares. If the field is larger than 5 hectares, then multiple composite samples should be collected. Large fields should be subdivided based on differences in soil type, crop history, and previous yields. Portions of a field with a history of high yields are unlikely to have a serious problem and need not be sampled intensively, whereas portions of the field with a history of low

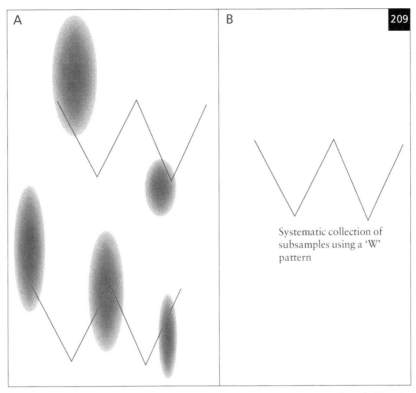

209 A scheme for sampling a large field prior to planting. **A**: Portion of the field with relatively poor yield the preceding season (shaded areas indicate areas of poor crop growth and yield due to high nematode population densities). Note that collecting two composite samples using a 'W' pattern for each will result in sub-samples from infested and noninfested portions of the field. **B**: Portion of the field with good yield the previous season, hence only one representative composite sample is collected.

yields should be sampled more intensively. Once the area to be sampled has been identified, sub-samples can be collected using a systematic sampling procedure. This can include any of several systems such as a zig-zag, 'W', or 'X' pattern (**209**).

When collecting samples from a current crop, it is still necessary to collect composite soil samples. One should avoid collecting sub-samples from plants that are dead, but instead concentrate on living plants that are exhibiting a range of symptoms (**210**). Separate samples should also be taken from around plants that appear to be healthy, for comparison. Sampling fields outside the cropping season or when the soil is very dry will yield very few active nematodes when extracted. In all cases, it is important to collect sub-samples from the root zone of the crop. Typically this means collecting the soil from beneath the crop canopy. If the crop is a

210 Collection of plant and soil samples from a soybean field with symptoms of severe damage due to nematodes, in this case from the sting nematode *Belonolaimus longicaudatus*. 'X' indicates appropriate locations for collecting plant and soil samples that represent a range of observed symptoms. Area for collecting samples from apparently healthy plants is also indicated (arrow).

perennial tree crop, then samples should be collected from near the drip line of the canopy and not adjacent to the trunk of the tree. The depth of sample taken will vary somewhat with the crop. The vertical distribution of nematodes is usually equal and proportional to the vertical distribution of the crop roots but it is seldom necessary to collect samples from depths of greater than 30 cm or from the upper 5 cm of soil which will contain few nematodes. The usual range is from a depth of 10–30 cm. Occasionally, more shallow samples are sufficient when plant roots grow near the surface.

Soil samples can be collected with any implements such as trowels, forks, hoes, narrow bladed spades or shovels, machetes, or large knives. However, they are collected most efficiently with sampling tools designed for the procedure, such as a standard Oakfield soil probe with a diameter of 2.5 cm. If using a shovel or spade, it is best to collect only a narrow column of soil from each shovelful of soil to avoid excessive sample volumes. The multiple sub-samples should be thoroughly mixed together in a large bag or bucket and a final sample of 1 to 2 litres of soil placed in an appropriately labeled plastic bag.

Collecting soil samples from beneath an existing crop will usually result in some root fragments also being collected. Nonetheless, it is best to also collect specific root samples to aid in the diagnosis. When doing so, one should dig up the plant so as to obtain as many of the fine feeder roots as possible. If the plant is pulled from the ground, then most of these feeder roots will be lost. If sampling a perennial crop, it is also important to collect feeder roots specifically from the current year's growth rather than larger and older roots. It will be difficult to make an accurate diagnosis from a sample that only contains large roots. Of course, some nematode parasites are rarely found in the soil or roots but are found primarily in the bulbs, corms, stems, or foliage. In such cases, care must be taken to collect the appropriate symptomatic tissues. Again, samples should not be taken from long dead plants as the parasites may be difficult to detect in such samples. It is best to collect samples from a number of live plants that are exhibiting a range of symptoms.

Collecting samples

- Collect a composite soil sample of 15–20 sub-samples
- A maximum of 5 hectare for each composite sample
- Concentrate on parts of field with history of low yields
- Use a systematic sampling pattern – zig-zag, 'X' or 'W'
- Collect from the root zone, under crop canopy
- Collect from living plants with a range of symptoms
- Collect from a depth of 10–30 cm
- For foliar nematode parasites, collect the appropriate plant tissues

Care of samples after collection

After the collection of plant and soil samples, it is critical that the biological vitality of the sample be preserved. The samples should be delivered or shipped to the diagnostic laboratory without delay. The samples should be protected from extremes of temperature. i.e. freezing or temperatures above 35°C. Thus, they should be packed in insulated containers and kept in a cool environment. Refrigeration (storage at 4°C) is not required if the sample is being processed within a day or two, but is helpful if the samples will be stored for longer time periods. It is usually not necessary to pack samples in ice for shipment, but shipping over a weekend or holiday period should be avoided. This will reduce the possibility of the samples being left unprotected on a loading dock or in a warehouse for several days. A good rule to follow is to treat the samples like perishable food that one wishes to consume in 3–4 days.

Care of samples after collection
• Deliver and treat samples without delay
• Protect from extremes of temperature
• Avoid shipping over a weekend or holiday
• Consider samples as perishable in 3–4 days

Extraction of nematodes from soil and plant samples

Typically, it is best to allow the diagnostic laboratory to extract nematodes from submitted samples and to identify the various parasites that may be present. However, it may be necessary to do the extraction prior to shipment or to have an initial examination of the nematodes. This can be true when the distances are great because shipping a large number of relatively heavy and bulky soil samples can be expensive. Additionally, regulatory concerns may prohibit the shipment of samples that may contain live pest species. Thus one may be required to extract and preserve the samples prior to shipment.

There are a large variety of methods for extracting nematodes from plant and soil samples (e.g. extraction trays, sieving, flotation, elutriation, centrifugation, flocculation) each with its own advantages and limitations. It is beyond the scope of this text to present a review of all of these methods; rather, a few of the simple, but effective procedures will be presented that can be used to extract vermiform nematodes and cysts from soil and plant tissues.

The Baermann-type methods

The Baermann funnel and its modifications are used to extract active nematodes from soil (or plant tissues). Basically, a coarse plastic or metal screen or sieve supports a layer of tissue paper on which is spread a known volume or weight of soil. This is placed in a suitable sized container (basin, plastic bowl, tray) and water is added until the soil or roots are just covered. The container is incubated for 24–48 hours (**211**). After this time, the screen, tissue paper, and soil are carefully removed leaving a clear water suspension of nematodes that have migrated through the plant tissue into the water. The nematodes can be concentrated by pouring the suspension through a sieve with pore diameters of 45–25 µm (325 to 500 mesh) and back washing from the sieve with a gentle stream of water into a small beaker. Alternatively, the contents of the funnel can be put into a large beaker, and allowed to sit for 1–2 hours, during which time the nematodes will settle to the bottom; then most of the excess water can be carefully decanted or siphoned off, taking care not to disturb the residue at the bottom of the beaker.

The actual Baermann funnel consists of a coarse mesh screen that fits just inside the upper lip of a large funnel (10–15 cm diameter) supported on a ring stand or a rack specifically designed to hold multiple funnels. A stopper or short length of plastic or rubber tubing attached to the stem of the funnel with a strong pinch clamp is used to seal the bottom of the funnel (**211**). The funnel is filled with tap water, then tissue paper is placed over the coarse mesh screen at the top of the funnel. A layer of soil or root fragments is then placed on the tissue paper. The water level should be adjusted until the water just covers the sample. The sample is incubated in the funnel for 24–48 hours at room temperature (it may be necessary to add more water after 24 hours to replace that lost due to evaporation), during which time the nematodes will migrate from the sample through the support tissue, and can be collected from the water in the funnel and concentrated as above.

The amount of soil that can be extracted with one funnel is dependent on the size of the funnel or other container that may be used. Regardless of the size and shape of the container, the thickness of the soil layer being extracted should not exceed 1 cm. Thus,

the larger the diameter of the funnel, the larger the size of the sample that can be extracted. Larger sample sizes will increase the reliability of the diagnosis.

The key to successful extraction of nematodes with Baermann-type methods are: (i) that the samples have been treated so as to preserve biological activity (dead nematodes will not migrate); (ii) to keep the sample immersed in water during the extraction period; and (iii) to have a shallow layer of sample and water to reduce the migration distance and to enhance diffusion of oxygen into the system. Because diffusion of oxygen

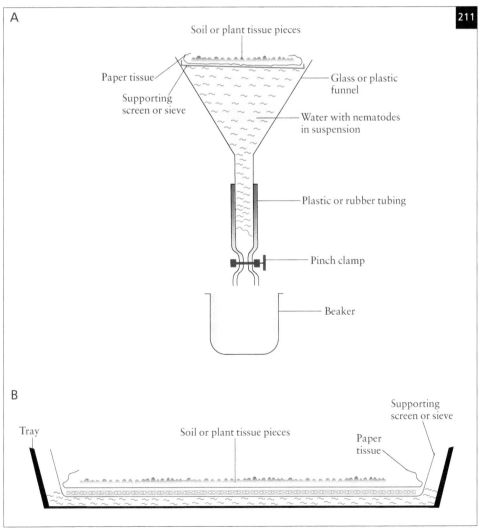

211 **A**: Baermann funnel; **B**: Tray modification for extraction of nematodes from soil and plant tissues.

into water is slow, reduced concentrations of oxygen in thicker layers of water will reduce the activity and migration of nematodes, which are aerobic organisms. Additionally, it is important to add the initial water and any replacement water added during the extraction period from beneath the sample to avoid washing any soil through the tissue. Also, pieces of tissue paper that extend beyond the funnel/container should be folded over the sample so that they will not act as a wick and drain water from the apparatus.

One of the limitations of the Baermann funnel is the relatively low extraction efficiency for very large nematodes such as *Xiphinema* and *Longidorus* spp. and for species that are relatively slow moving such as *Criconemoides*. These nematodes can be extracted by the sieving method described below. Additionally, it will not permit estimation of numbers of cysts in soil (only pre-parasitic J2 stages and adult males can be recovered) or sedentary, swollen stages of endoparasites such as *Globodera*, *Heterodera*, *Meloidogyne*, *Nacobbus*, *Rotylenchulus*, or *Tylenchulus* spp.

Plant tissues

Nematodes can be extracted from roots and other plant tissues (corms, bulbs, rhizomes, stems, leaves, seeds) using the Baermann-type methods described above. Normally, a known weight of roots is taken to determine populations per gram of tissue. The roots should be chopped into small pieces or macerated in a blender to release the nematodes. Plant tissues should be left extracting for at least 48 hours. It is also possible to extract nematodes from plant tissues by simply incubating them in a dish of water for 3–4 days. In some cases, nematodes in plant tissues can be observed directly under a stereoscopic microscope by teasing the tissues with a small knife or needle in a dish of water.

Sieving

The sieving extraction method is used to extract inactive or sluggish nematodes. A suspension of soil in water is poured through a series of sieves of different sizes depending on the nematodes likely to be present. Most plant parasitic nematodes can be collected on sieves of mesh aperture sizes 90 μm and 53 μm. These are arranged in a bank with a larger mesh size (2 mm) on top to retain stones and other debris. A known volume of soil (200–400 cm^3) is placed into a large bucket of water. The water and soil is then stirred vigorously and left for 30 seconds before being poured through the bank of sieves that have been moistened. Nematodes collected on the smaller aperture size sieves are then washed into a beaker with a gentle jet of water applied to the *back* of each sieve. The process is repeated three times. The nematode suspension in the beaker is left for a number of hours and concentrated as before by decanting or siphoning off the excess water.

Cyst extraction

The rounded dead females of *Globodera* and *Heterodera* species, the cysts, occur free in the soil and, because they float when dry, they can be easily extracted from the soil after it is dried. A small amount of dried soil, 50–100 cm^3, is put into a narrow 1–2 litre beaker or conical flask to which some water is added, and the container is then shaken or stirred vigorously. Following this, the container is filled with water to the brim and left to stand for a short period. If cysts are present in the soil, they will be seen floating to the surface together with plant debris. These cysts can be removed manually with a pipette or fine brush or gently tipped onto a filter paper.

Staining nematodes in plant tissues

Nematodes are translucent organisms and difficult to see microscopically when embedded in plant tissues. To make them visible they can be stained by washing the roots to remove soil and debris and boiling for 3 minutes in a solution of equal parts lactic acid, glycerol, and distilled water + 0.05% acid fuchsin or cotton blue stain. The stain is then cleared in a 50:50 mixture of glycerol and distilled water. Nematodes should be stained bright pink or blue, depending on the stain used.

Preservation of extracted nematodes

Nematodes are best identified from live specimens. Extracted nematodes in water can be stored for 1–2 weeks in a refrigerator. Living nematodes can also be shipped at ambient temperatures if they will arrive at their destination within 2–3 days. Otherwise, the samples must be preserved. Unless appropriate permits are obtained, all international and many intra-national shipments must be of dead, preserved specimens. Nematodes need to be heat treated for preservation as heating nematodes to 55–60°C kills them and relaxes them into characteristic shapes.

Nematodes can be readily preserved in 2% formaldehyde. A quick and reliable method is to concentrate the sample to a volume of 5–20 ml, then add an equal volume of boiling 4% formaldehyde to the sample that is at room temperature. Because the formaldehyde fumes are toxic, one must take appropriate safety measures to avoid inhalation of the fumes. After the sample has cooled, it can be transferred to a small vial or bottle, sealed to prevent loss of fluid, then packaged for shipment. Nematodes preserved in this manner can be stored at room temperature for many months with little loss of integrity.

Nematode identification

The identification of nematodes is primarily based on morphological characters and morphometric measurements. Because males are absent or rare for many species, identification is most often based on the mature female. Characteristics of juvenile stages may be useful in identification of some genera and species.

Most plant parasitic nematodes are microscopic and are not visible to the naked eye. Some morphological traits are readily observed using the low magnification (30–60×) of a stereoscopic microscope (**212**). These include the overall size and shape of the nematode, presence and shape of the stylet and stylet knobs (all plant parasites have a stylet but not all stylet-bearing nematodes are plant parasites), stage of development (juvenile vs. adult – only in the adult is the vulva of the female present, absence of this character indicates that the specimen is a juvenile), position of the vulva and tail shape (rounded, conical, or filiform). Other morphological characters can only be seen at higher magnification with a compound microscope.

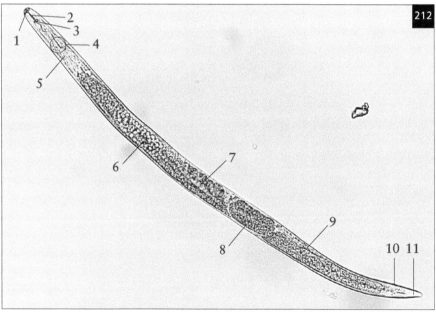

212 Photomicrograph of *Pratylenchus* sp. with the main morphological features indicated. 1 head and lip region; 2 stylet; 3 stylet knob; 4 median bulb; 5 oesophagus; 6 intestine; 7 ovary; 8 egg; 9 vulva; 10 anus; 11 tail.

Bibliography

Barker, K. R., Carter, C. C., and Sasser, J. N. (eds) (1985). *An Advanced Treatise on Meloidogyne: Vol. II. Methodology.* North Carolina State University Press, Raleigh, US.

Barker, K. R., Pederson, G. A., and Windham, G. L. (eds) (1998). *Plant Nematode Interactions.* American Society of Agronomy, Madison, US.

Brown, R. H. and Kerry, B. R. (eds) (1987). *Principles and Practice of Nematode Control in Crops.* Academic Press, Sydney, Australia.

Chen, Z. X., Chen, S. Y., and Dickson, D. W. (eds) (2004). *Nematology: Advances and Perspectives. Vol. I. Nematode Morphology, Physiology and Ecology.* Tsing University Press, China; CABI Publishing, Wallingford, UK.

Chen, Z. X., Chen, S. Y., and Dickson, D. W. (eds) (2004). *Nematology: Advances and Perspectives. Vol. II. Nematode Management and Utilization.* Tsing University Press, China; CABI Publishing, Wallingford, UK.

Desaeger, J., Rao, M. R., and Bridge, J. (2004). Nematodes and other soilborne pathogens in agroforestry. In: *Below-ground Interactions in Tropical Agroecosystems: Concepts and Models with Multiple Plant Components.* Noordwijk, M., Cadisch, G., and Ong, C. K. (eds). CABI Publishing: Wallingford, UK, pp. 263–283.

Evans, K., Trudgill, D. L., and Webster, J. M. (eds) (1993). *Plant Parasitic Nematodes in Temperate Agriculture.* CABI Publishing, Wallingford, UK.

Ferraz, L. C. C. B. and Brown, D. J. F. (2002). *An Introduction to Nematodes: Plant Nematology.* Pensoft Publishers, Sofia, Moscow.

Jepson, S. B. (1987). *Identification of Root-knot Nematodes (*Meloidogyne *species).* CABI Publishing, Wallingford, UK.

Luc, M., Sikora, R. A., and Bridge, J. (eds) (2005). *Plant Parasitic Nematodes in Subtropical and Tropical Agriculture.* 2nd edn. CABI Publishing, Wallingford, UK.

Mai, W. F. and Mullin, P. G. (1996). *Plant-parasitic Nematodes: A Pictorial Key to Genera.* 5th edn. Comstock Publishing, Ithaca, US.

Marks, R. J. and Brodie, B. B. (eds) (1998). *Potato Cyst Nematodes: Biology, Distribution and Control.* CABI Publishing, Wallingford, UK.

Nickle, W. R. (ed) (1991). *Manual of Agricultural Nematology.* Marcel Dekker, New York, US.

Sasser, J. N. and Carter, C. C. (eds) (1985). *An Advanced Treatise on Meloidogyne. Vol. I. Biology and Control.* North Carolina State University Press, Raleigh, US.

Schmitt, D. P., Wrather, J. A., and Riggs, R. D. (2004). *Biology and Management of Soybean Cyst Nematode.* 2nd edn. Schmitt & Associates of Marceline, Marceline, US.

Siddiqi, M. R. (2002). *Tylenchida: Parasites of Plants and Insects.* 2nd edn. CABI Publishing, Wallingford, UK.

Starr, J. L., Cook, R., and Bridge, J. (eds) (2002). *Plant Resistance to Parasitic Nematodes.* CABI Publishing, Wallingford, UK.

Waller, J. M., Lenné, J. M., and Waller, S. J. (eds) (2002). Plant Pathologist's Pocketbook. 3rd edn. CABI Publishing, Wallingford, UK.

Wharton, D. A. (1986). *A Functional Biology of Nematodes.* Johns Hopkins University Press, Baltimore, US.

Whitehead, A. G. (1998). *Plant Nematode Control.* CABI Publishing, Wallingford, UK.

Glossary

Aggressiveness – the relative ability of a parasitic nematode to cause damage to the host

Anhydrobiosis – condition during which some nematode species are able to undergo a loss of water and assume a coiled, shrunken posture, associated with survival of long periods of drought

Baermann funnel – an apparatus used to extract nematodes from plant tissues or soil

Biological control – reduction of nematode populations by the activity of another organism, typically a fungus or bacterium, that parasitizes or is antagonistic to the nematode

Burrowing nematodes – *Radopholus* spp.

Carbamate nematicides – a class of nematicides that also have insecticidal activity and some systemic movement in plants; examples include aldicarb and oxamyl

Citrus nematode – *Tylenchulus semipenetrans*

Control – tactics used to reduce nematode population densities and increase crop yields

Crop rotation – rotating crops in a sequential cropping system; the aim of any rotation for nematode management is to allow a sufficient interval to elapse after a susceptible crop so that nematode populations can decline to a level that will allow the next susceptible crop to yield at an acceptable level

Cuticle – the exoskeleton of nematodes

Cyst – the body of a dead female cyst nematode that protects the dormant eggs contained, characterized by a darkly pigmented cuticle and important for survival

Cyst nematodes – nematodes that form cysts, mainly *Globodera* and *Heterodera* spp.

Dagger nematodes – *Xiphinema* spp.

Dauer larva – an alternative stage of development for some nematode genera, specifically adapted for dispersal and survival of adverse conditions

Dormant – individual or life stage in an arrested stage of development and with greater potential for survival of adverse environmental conditions

Ectoparasites – nematodes that feed primarily on the surface of the host plant and do not penetrate into the inner tissues of the root, tubers, or foliage

Egg mass – a gelatinous matrix produced by rectal glands of females of several genera of sedentary parasites and into which eggs are deposited

Endoparasites – nematodes that feed primarily on the inner tissues of a host, such as the root cortex or leaf parenchyma; some species are migratory, whereas others are sedentary

False root knot – *Nacobbus* spp.; nematodes causing the formation of root galls on susceptible hosts similar in appearance to those cased by true root knot nematodes

Foliar nematodes – nematodes feeding on stems, leaves, inflorescences, and seed (apart from peanut)

Fumigant nematicides – volatile compounds that are toxic to nematodes and other soil organisms, applied in such a manner so as to creat a protected zone in the soil into which susceptible crops can be planted; typically these are phytotoxic and move through the soil as a gas

Galls – localized swelling of roots or other plant organs due to hyperplasia and hypertrophy caused by parasitism by some nematode genera

Giant cells – specialized host cells induced by nematode parasites that serve as nurse cells for the parasite; cells are characterized by a dense cytoplasm, absence of a central vacuole, multinucleate, and derived from a single cell by repeated mitosis without cell division; root knot nematodes induce 3–5 giant cells at each infection site and feed exclusively from these specialized cells

Gonad – ovary of female, testis of male, fully formed only in adults

Green manure – fresh plant shoots that are incorporated into the soil to increase organic matter content of soil and to stimulate microbial activity in the soil, often useful in effecting a reduction in populations of parasitic nematodes

Habitat – the natural environment of an organism, the locality where an organism lives

Host – any plant species on which the nematode can complete its life cycle; good hosts support high levels of reproduction whereas poor hosts support relatively low levels of reproduction

Infected – the establishment of an intimate relationship between the parasite and its hosts, characterized by an exchange of metabolites (also infection)

Infested – the physical association of a parasite with some location, such as soil or plant surface but lacking the biological characteristics of an infection (also infestation)

Intolerant – when the growth of a plant host is reduced as a result of the parasitism by a nematode, as opposed to a tolerant plant species, the growth of which is relatively unaffected by nematode parasitism

Juvenile – immature stages (typically four in number, J1, J2, J3, J4) of nematodes, the preferred term to larva

Lance nematodes – *Hoplolaimus* spp.

Larva – immature stage of insects (the preferred term for nematodes is juvenile)

Lesion – localized area of disease, typically characterized by host cell death and dark discoloration of affected cells

Lesion nematodes – nematodes that cause necrotic lesions in host roots and other infected tissues as a result of parasitism, *Pratylenchus* spp.

Management – strategies used to suppress nematode population densities and maintain them at densities at which there is little or no crop loss

Median bulb – muscular pump of Aphlenchida and Tylenchida nematodes for withdrawing contents from host cells, located in the esophagus, also called the metacarpus

Migratory – nematodes that retain ability for locomotion on and in plant tissues and soil throughout their life cycle

Necrosis – localized area of cell death, typically resulting from pathogenesis, often characterized by purple, brown, or black colour

Needle nematodes – *Longidoros* and *Paralongidorus* spp.

Nematicides – natural or synthetic chemicals that are toxic to nematodes; two major classes are fumigants and nonfumigants

Nematodes – invertebrate organisms, triploblastic, pseudocoelomic, bilaterally symmetrical, mostly aquatic or soil inhabiting, many parasitic on other organisms including mammals, fish, arthropods, and plants

Nematode wool – an aggregation of nematodes, typically while entering an anhydrobiotic condition, a survival mechanism

Nonfumigant nematicides – nematicides that are liquid or granular formulations, dissolve in soil water when applied to soil and move through the soil in the water phase

Oesophagus – anterior portion of the nematode from the head to the juncture with the intestine

Organophosphate nematicides – a class of nematicides that also have insecticidal activity and some systemic movement in plants; an example is phenamiphos

Ovary – female reproductive organ that produces eggs, one or two in number depending on the genus, fully formed only in the adult stage

Parasite – an organism that derives its nutrients from another, typically larger, organism; characterized by long-term intimate associations (see infected); often results in disease

Parasitism – act of being a parasite

Parthenogenesis – reproduction without males

Pathogen – an agent that causes disease

Pathogenesis – the act of causing disease

Pathotype – subspecific variation within a species distinguished by the ability to parasitize plant hosts having resistance to other genotypes of that nematode species

Race – synonymous with pathotype

Red ring nematode – *Bursaphelenchus cocophilus* (previously *Rhadinaphelenchus cocophilus*), causes the red ring disease of coconut and oil palm, characterized by red discolouration of vascular tissues of infected stems to give the red ring symptom

Resistance – the ability of a plant to suppress nematode reproduction relative to reproduction on a susceptible plant genotype that supports relatively high levels of reproduction; resistant plants usually have greater yields in nematode infested fields than susceptible plants

Ring nematodes – currently *Criconemoides* spp. (formerly also known as *Criconemella* spp., *Macropostonia* spp., and *Mesocriconema* spp.)

Root knot nematodes – *Meloidogyne* spp., nematodes that induce the formation of galls (knots) on roots and other plant tissue at the site of infection

Sedentary – nematodes that lose locomotion abilities as they mature, typically endoparasites or semi-endoparasites with greatly swollen bodies at maturity

Semi-endoparasites – nematodes that are at least partially embedded in host tissues during parasitism and feed on internal host cells, typically sedentary; examples are citrus nematode (*Tylenchulus* sp.), cyst nematodes (mainly *Globodera* and *Heterodera* spp.), and reniform nematode (*Rotylenchulus* sp.)

Spiral nematodes – *Helicotylenchus* and *Rotylenchus* spp.

Stem and bulb nematodes – *Ditylenchus dipsaci*

Sting nematode – *Belonolaimus* spp.

Stubby root nematodes – *Paratrichodorus* and *Trichodorus* spp.

Stylet – hollow, needle-like organ, formed in the stoma of nematodes and used to pierce host cells and remove host cell contents

Sub-tropical climate – areas outside the true tropics but which are characterized by warm temperatures, with mild winters and few, if any, days with temperatures below freezing (0°C)

Susceptible – a plant species or genotype that supports a relatively high level of nematode reproduction

Syncytia – specialized host cells induced by nematode parasites, which serve as nurse cells for the parasite; similar to giant cells but are formed by the coalescence of several adjacent host cells to form a multinucleate cell

Tail – that portion of the nematode body from the anus (female) or cloaca (male) to the posterior terminus

Temperate climate – climates similar to that which exists outside of the tropics, usually with distinct winter periods during which temperatures are often below freezing (0°C)

Tolerant – a host that suffers relatively little suppression of growth due to nematode parasitism in comparison with other genotypes of that host species that are intolerant and suffer relatively large suppression of growth due to parasitism

Trap crop – crop planted and invaded by sedentary endoparasitic or semi-endoparasitic nematodes, after which the crop is destroyed prior to the nematode completing its life cycle, thus reducing nematode populations that may attack a subsequent susceptible crop

Tropical climate – climates similar to that existing in the tropics, that region between the Tropic of Cancer and Tropic of Capricorn; characterized by warm temperatures year round, often humid with excessive rainfall

Vector – an organism that transmits or disseminates pathogens and parasites from one host to another

Vermiform – having a worm-like shape

Virulence – referring to the ability of nematode pathotypes or races to parastize host genotypes that are resistant to other populations of that nematode species

Vulva – female orifice leading to uterus and ovaries

Index